The Gospel of Jesus Green

The Gospel of Jesus Green

Home for All, Not Just for Humans

NEIL J. WHITEHOUSE

WIPF & STOCK · Eugene, Oregon

THE GOSPEL OF JESUS GREEN
Home for All, Not Just for Humans

Copyright © 2024 Neil J. Whitehouse. All rights reserved. Except for brief quotations in critical publications or reviews, no part of this book may be reproduced in any manner without prior written permission from the publisher. Write: Permissions, Wipf and Stock Publishers, 199 W. 8th Ave., Suite 3, Eugene, OR 97401.

Wipf & Stock
An Imprint of Wipf and Stock Publishers
199 W. 8th Ave., Suite 3
Eugene, OR 97401

www.wipfandstock.com

PAPERBACK ISBN: 979-8-3852-0024-5
HARDCOVER ISBN: 979-8-3852-0025-2
EBOOK ISBN: 979-8-3852-0026-9

06/10/24

The author gratefully acknowledges copyright permission from:

GreenBiz Group Inc., for blog quotes by Joel Makover.

Hal Leonard LLC, for use of the lyrics of "You've Got to Be Carefully Taught" from *South Pacific*. Lyrics by Oscar Hammerstein II. Music by Richard Rodgers. Copyright © 1949 Williamson Music Company c/o Concord Music Publishing. Copyright renewed. All rights reserved. Used by permission. Reprinted by permission of Hal Leonard LLC.

Hope Publishing, for use of the hymn "The Church of Christ in Every Age," words by Fred Pratt Green. Copyright © 1971 Hope Publishing Company, www.hopepublishing.com. All rights reserved. Used by permission.

Oxford Publishing Limited, for use of "A Dream of the Cross," in *Earliest English Christian Poetry: Translated into Alliterative Verse, with Critical Commentary*, by Charles W. Kennedy. Copyright © 1952 Hollis & Carter. Reproduced with permissions of the Licensor through PLSclear.

Venny Vega-Cárdenas at IORN, for the quotation from "Nature as a Subject of Rights" (The Atrato River Case).

Images from Wikimedia Commons are courtesy of Creative Commons, https://creativecommons.org/licenses/by-sa/4.0/deed.en.

Scripture quotations are taken from the New Revised Standard Version, copyright @ 1989, Division of Christian Education of the National Council of the Churches of Christ in the United States in the United States of America. Used by permission. All rights reserved.

Every effort has been made to respect copyright norms. We would be glad to hear from any person or publisher who questions the use of their material.

To John A. Cooke,
dear friend and generous host

Contents

Acknowledgments | ix

Introduction | 1

Chapter 1
Which Jesus? | 16

Chapter 2
What Do You Mean, Green? | 38

Chapter 3
The Green Jesus? | 58

Chapter 4
Deeper Green | 77

Chapter 5
Jesus Christ Is Jesus Green | 109

Chapter 6
A Passion for Green | 132

Chapter 7
How Green Is Our Savior | 158

All-Age Epilogue: The Court of Nature | 191

Bibliography | 195

Index | 207

Acknowledgments

The orientation of this book for secular as much as religious readers reflects my Methodist mentors like Rev. Lord Donald Soper, who stood at Speaker's Corner, Hyde Park, London, for decades and spoke with the conviction that the Jesus story has relevance in every generation.

Many friends have been crucial to this publication. Sincere thanks to Paula Kline and Patrick Barnard for their belief and advice in my writing; to Robert Taylor for the author photo; Jason Prince for manuscript criticism; Alison Webster for publishing advice, networks, and knowledge of the writing world; and François Joly for photographic assistance.

Thanks for your unique help: Nikita Eaton-Lusignan, Tristram Wyatt, Michael Arditti, Giles Goddard, Bernard Lynch, Karen Oliveto, David Horrell, Steve Plant, Jill Whitehouse, John A. Cooke, Alexander Lavoie, and Nigel Penney: the butterfly effect means every gesture counts, thank you! Thanks also to Matt Wimer and the team at Wipf and Stock Publishers, for their diligence, care, and patience in bringing this book to reality.

In an era when teaching and education are undervalued, I wish to thank Woodfield Primary and Junior School; Wolverhampton Grammar School; the Faculty of Zoology, Liverpool University; Wesley House, Fitzwilliam College, and the Divinity School of Cambridge University; McGill Ecumenical Chaplaincy (now St. Martha's Chapel) and the Student Services of McGill University.

Lastly, I thank and salute the board and members of Westmount Park United Church, who warmly embraced the Green orientation of my ministry among them, and gave me space to write and share some discoveries.

Most of my working life as a church minister is dealing with invisible things, like human relationships. Most of the tensions of my clients on a massage table cannot be seen either. At least to write leaves a trace in the sand. The slower, reflected pace of a book allows for the deeper change I hope to encourage. It explores one of the most important subjects of our century: how we become a species whose actions correspond with our knowledge of the living world.

The original Gospels dealt with existential crises in very different times. Jesus Green offers a gospel that disturbs us to *live* differently; for us to come to ourselves and make home for all.

Neil J. Whitehouse, Montreal, February 1, 2024

Introduction

WONDERFUL LIFE

Love in the Time of Cholera by Gabriel García Márquez begins with the lyrical death of Dr. Juvenal Urbino, who falls from a ladder trying to recapture his parrot.[1] I had to be careful not to mimic him at my special place of wildness in Montreal, when life was stranger than fiction.

Montreal is so large you rarely sense it is an island, but going to a smaller one helps. In the middle of the St. Lawrence Seaway is Isle St. Hélène. It was extended in the sixties for the iconic Expo '67 and has a great aquatic center. Clubbers also enjoy it for massive outdoor dance parties, but nature on the original island hangs tough, too, happily ignored by the throngs. A turreted stone tower built in 1937 peaks the skyline. Close by is a small balcony that once gave a view back to the city. Now you just see trees. I go there by biking over the dramatic arc of the massive Jacques Cartier Bridge. It is a stunning contrast to whiz down the cycle path and then turn off, sharp right, into the calm of forest bursting with bird song. Often birds come close: warblers, goldfinch, woodpeckers, nuthatch, waxwing, kingbirds; further away blue herons and egrets fly lazily to new roosts. Pairs of turkey vultures circle, on watch for carrion, or hang out on the stone tower as Gothic statues.

One hot Friday afternoon, I swam a little in the outdoor pool, then went to my wild place. Being there, doing nothing, is my thing. The air was humid, still and luxurious. As I drank it in, magic happened.

Magic makes you doubt your eyes: I moved to Montreal in 2001 and had never seen a hummingbird, except on vacation in Cuba and a dead one on McGill University campus. But that day, there, close to head height, one zip-zapped in bejeweled green glory, searching in the hyper way hummingbirds do. I blurted, "Look, a hummingbird," to an oblivious passer-by, and as soon as it had arrived, the bird was gone. A grateful wonder flowed into me until I was distracted again. On the ground, twenty feet away, was

---

1. Márquez, *Love in the Time*, 45.

a moving banana. No, another bird; not a goldfinch, a little bigger, and this was the fiction moment: a budgie. Another twenty-year first. I had a flash to a canary named Zacharie, which we lost to the outside world, so my heart went out to the bird and its carer.

I went closer, expecting to scare it, but this bird was tired and hungry. The next hour I chased it. Well, I approached it, slowly, grabbed in hope and swore in futility. My failure shocked me. It seemed ready for salvation. How could I not succeed? Five attempts later, each time with less conviction, I realized it wasn't me being clumsy. This was difficult to do without a net. The bird's survival instincts would be the last to die. The paradox that a natural fear was a curse for this bird made me upset. But as I perched on a ledge, I remembered Dr. Urbino, and maybe this saved me from broken bones. The sixth grab drove the bird across treetops. Goodbye.

Back home, in the urban realities of Plateau Montreal, I was disturbed. What had just happened? How can two events with birds, lifetime firsts, happen like birthing twins? Why had I been so concerned to save the budgie? I had no care for the hummingbird, whose life is normally much shorter and more precarious. The answer shouted out the gospel of Jesus Green. The hummingbird was at home, and the budgie was homeless. It was incapable of thriving and dependent on human care.

My struggle had been to bring the budgie home. Being at home was key. I felt bad about failing to catch this bird. The best I can do is share the story with you, and how it teaches the wisdom, needed to care for some living beings, but also that others thrive very well without us, if we leave them be. One bird is not the other, and we are obliged to know the difference.

I look into myself, my history, and swim in this wild world with wonder. It seems vital to learn from hummingbird and budgie, to let the truth of life change human history. It is wonderful but also demanding. We are alive at an electric time of human choice and possibility. Literally. Technology has created great perils but also new consciousness of what is truly precious. We are so powerful in many ways and inadequate in others. To wonder permits us to step away from our arrogance and to accept our place in things.

This is how I can dare to suggest renaming Jesus "Green." It is an invitation to participate in an environmental Copernican revolution—something new and life changing about being human that is as wonderful as a hummingbird. This rejects the assumption that science and religion are like oil and water. Many Christians still put religious beliefs before scientific evidence. Many scientists have long given up on religion offering any rational worldview, but I am lucky to have had great teachers who knew otherwise, at Liverpool University for zoology and Cambridge University for theology.

This matters, not simply because millions of people are religious and need to integrate science with their beliefs. It matters because an essential part of the paradigm shift to sustainable living is the soul within human cultures and systems. Essential meaning is intangible as well as necessary, like the experience of being in a crowd that is enthralled by music or sport. This is the content of how to move from despair to wonder. It is the imponderable in the process of most democratic election campaigns. I am not advocating any particular religious practice but the admission that wonder remains a human need and tendency, so that Jesus Green can prophesy a fuller sense of being alive, being human.

If our descendants do not take life for granted, they will notice hummingbirds.

When news headlines tell of new human or environmental violence, they cast shadows on my wonder for the world. The shadows go back more than a century. Yet something is new, something that justifies Jesus Green. When Charles Dickens exposed the waste of life bound up with the development of British society at the height of the Victorian era, he gave us the orphan Oliver, who asked the master, "Please, sir, can I have some more?"[2] Today, Oliver is a child refugee, homeless because of a civil war brought on by drought and failed harvests. From a century that gave us answers to *The Origins of Species*, we live in one of climate instability, mass extinctions, and a new and overwhelming question:

How can we live in a civilized manner without consuming the planet that gave us life?

This question is behind all Green concerns and provokes a green-blue paradox: our wonder at the natural world provokes lament. Think William Wordsworth, musing on a daffodil; Gerald Manley Hopkins on a kestrel; Walt Whitman on *Leaves of Grass*; Alice Walker on "Dear God. Dear stars, trees, dear sky. dear peoples. Dear Everything. Dear God."[3] Now we face the "dear" David Attenborough, whose lifelong career of natural history broadcasting makes him such a world authority; we know him as a truthful man, presenting the planet Earth in peril. In his lifetime, the crisis has become unavoidable.

I admit there is work to do: organized religion has a way of mothballing crises, especially Christianity in the West, which is so loaded with social history. But can the authentic wonder you know for the world allow you to change what you think about Jesus? Will you suspend any preconceptions that this book is just a nice idea? We face a planetary drama of

---

2. Dickens, *Oliver Twist*, 13.
3. Walker, *Color Purple*, 242.

overpopulation and the Hollywood dream. You and I and all our ways of shopping, making, growing, selling, governing, need to be reshaped truer to this wondrous life on earth. How can we change? Is there power to change in the Jesus story that still makes sense? Does this give people hope to live differently?

It may seem trite and superficial to propose a gospel of Jesus Green. It comes from left field, strange and irrelevant to the urgent tasks: wishful thinking from a pastor. Yet something of the global scale of religion is what is needed; a movement of the spirit as well as hands and feet; an imagination that connects to the creative capacities within each of us, for a flourishing earth.

This has a necessary political realism too. How do people choose the way they vote? Explain to me the rise of populist leaders. Is it the unhappiness of millions in need of hope, against a threat to life as they know it, and who grasp for any promise to stop them drowning? Popularism in the twenty-first century can be the root of evil, as much as the love of money! We need a vocabulary and sufficient consensus that is neither a lie nor without hope: true at a deep level of human need and natural reality: words and shared understandings with soul, to face failure with a resolute nevertheless. It is bound to be multifaceted, new and old at the same time.

To reach that point, we will discover a creative tension between an elusive Jesus and the world we know. This takes us beyond religion: a tension that allows the inheritance of Jesus to be experienced within and between us.

But I doubt these claims strike you deeply, not yet. Why should they? Test, reflect, and discover through the world I describe and the world you know. Let this "to and fro" be the process of Jesus Green, a gospel of home for all in a wonderful world.

PEOPLE AND PLACE—ORDINARY CHURCH

I am no Dickens, and neither do I wish to embarrass my congregation with too much information, but to explore Jesus Green, I will tell you about my church. It tells you a little where I am coming from, as a certain sort of Christian. Our being and identity come through people and place.

Each Sunday afternoon a few people come to worship at Westmount Park United Church. People trickle in, making it difficult to begin without feeling underwhelmed. We can be just fifteen. Twenty-five is a good day. (I leave to one side the strangeness of hybrid worship we offer, since COVID-19.) Fewer than five attendees earn a wage. Thankfully we have a great building, and the community has long relied upon rental income from tenants who run programs for children, seniors, and the arts. The sanctuary was built in 1929 by a notable architect Alexander Perry in neo-Gothic style. It is the third church on the site bought by Methodists in 1889 and built to accommodate merging congregations with a sense of growth and importance for the church at the heart of society. This mood continued into the 1960s, and few anticipated the drastic decline since.

One "Saturday night–Sunday morning" experience at church told me a lot: A film club rented the sanctuary to show silent black-and-white movies, like the first *Dracula* or *Hunchback of Notre Dame*, with live musicians playing unscored to what they see on screen. New to the church, I was amazed how this packed people in, of all ages, from across the metropolis. I still have my selfie of that first Saturday night, with my caption, "Look! Week four of my ministry . . ." Joking aside, it was startling proof Sunday worship is not what people want. Not just now.

Imagine however, that this is to miss out on something. Why bother to worship, given the small numbers and seeming failure of the cause? Imagine that we can step through the dead ends of our assumptions about church and church language and forgive the church for bad practice, so that a worship service can come alive. Imagine it can resonate with the power of confrontation between Jesus and his contemporaries and carry the sort of joy and love that gave hope to his followers.

Compare this with something you participate in like a hockey game or soccer match, or the yoga class or theater performance or the rave with a famous DJ. Once you go regularly, what makes an experience of any of these stand out? How many boring soccer matches have I watched? None of the events we support are *inevitably* out of this world, but that extraordinary move, the exhilarating release of tension, a sense of reality in the make-believe of theater, an ecstasy of body-mind connection in dance; whatever that astounding "high" is, comes only out of comparing it to the more mundane and how things can come together, unexpectedly, to make it special. You wait for it to happen because you know it has that potential.

I can't take you to church, but even if I could, it is likely to be the same underwhelming experience.

Instead, consider the people who still attend. Suspend your awkwardness and come with me to be in our place. This is when it gets interesting because, trust me, the diversity of a group of silver-haired people is extraordinary. Sprinkled among them, the "youngsters" are equally unique and distinct from their peers. I don't think we are all stupid, dependent, lonely, confused, or inadequate—these being some of the reasons we might be attached to a long-lost cause (though I can be all these things at certain times).

Take a moment in worship at Westmount Park: holy communion. It is my adaptation of receiving the sacrament at St. James, Piccadilly, Anglican Church, London, in the early nineties; the era of Rev. Donald Reeves. I found it helps to bring out the power of the ritual. Instead of people receiving bread and wine in the form of a lineup, I invite our flock to come forward as a group, or holy huddle, around the communion table. This is when we can seem very fragile, as older people move slowly, some are hesitant, and some move with pain. To come forward is not what people grew up with; it will always be strange, to some extent challenging even for those who appreciate it. Like the times I find it a challenge to get out of bed in the morning. But the effort is worth it: you never know who you will be "elbow to elbow" with, to receive something holy. You receive as an individual, yet you are forced to appreciate other people. Closer together we sense the reality of being a group, rather than pious individuals.

Is it the start of what keeps us going as Christians? Me to be me, but with you, to be us. Jesus' command "Love one another!" is not so abstract.

The small numbers do not mean small experiences. We experience what lies behind the New Creed of The United Church of Canada, at least the first line: "We are not alone..."[4]

Simple words, but the truth of it is deep. It renews. Being together in this place makes a connection with millions of others in the past and across the globe.

Now consider the place in which we meet and how it gives a sense of the present moment for worship. The slogan *church is people, not the steeple*, sidesteps the realities of our being material animals, and you know this from the holy places of your past.

Alexander Perry designed our stone-clad church to give a sense of a longer history than a 1929 build, with a tall bell tower and high vaulted ceiling. Dark wood pews originally seated up to three hundred people with a central aisle and two transepts. This gives the sanctuary a classic cross shape and directs the gaze to the front with font, pulpit, choir stalls, and organ. It fits the identity of a church of wealthy and influential Westmounters, who remained modest about their means.

The Gothic-style sanctuary is perfect for those black-and-white movies; perfect also for ten stained glass windows by a remarkable Montreal artist, Charles W. Kelsey. The scenes in the windows express a humanity and a love for nature and Bible stories. They remember families of the church and those who served in two world wars.

---

4. United Church of Canada, "New Creed," line 1.

Music is a central part of the worship experience, to explore emotions and connect us to one another. As a good Methodist I find hymns add a final touch, so when the words and the music fit together, frissons start to happen. The Casavant organ is perfectly sized for the space to end worship in stunning fashion, with a postlude by Bach!

For visitors present for a wedding or funeral, there are usually moments when something in this rich mix of religious activity delivers. The special reading, read nervously by a family member about love, strikes home, and you know many people are reaching for a tissue because of its beauty and truth. The struggle of a grieving daughter to name her father's mixed-up way of loving her family: honesty, anger, and love combine. Far more than what must be done at a funeral, it is cathartic, healing, changing. These things are increasingly happening outside of church, of course, but in the sanctuary, the place helps us to acknowledge our personal milestones are part of the greater stream of human history as each generation faces fundamental values and questions of existence. This is the religious power.

I know that this is out of date, but regular Sunday worship, the sort people have left in droves, still moves and nourishes. The sense of transcendence, being lifted out of ordinary to something expansive, awe inspiring in limitless scale and joy, is commonplace. Cosmic awareness also has a sense of closeness to something special, not human; that somehow in this place, this moment, huge energy and goodness are present—a sense of immanence. To travel between these experiences of transcendence and immanence beats any roller-coaster ride because it enhances how we see and interpret life itself. Grubbiness dealt with in sermons speaks of what is holy. It gives us the ability to face our dark side as well as our goodness and success.

None of these aspects insult your intellect, your proper sense of truth, and your knowledge of many other things.

I see these sorts of religious highs expressed unconsciously in superhero movies, but the frissons in church are in our world, not by an escape to others. It comes back to the elbow-to-elbow experience of being church; church as a meeting between people who over time, and from time to time, discover the courage to be honest together.

Meanwhile, the church you know may have been very different than my descriptions—a contradiction to truth or care, or both. I allow for the church to be a place for "sinners." It fails as an institution. But Jesus Green is new. Jesus Green offers a renewal of church as well as a challenge to secular life.

The common ground of people and place remains key for a twenty-first-century "salvation."

HONORING THE *TITANIC*

The sanctuary of Westmount Park United Church has ten magnificent stained glass windows, and one stands out as special for Jesus Green. The Allison window is dedicated to the wealthy businessman George Allison, but it is the story of his nephew Hudson (Trevor) that gets our attention.

Hudson was on board RMS *Titanic* on April 15, 1912, but was saved thanks to his nanny, Alison Cleaver. He was only eleven months old and traveled with his parents, Hudson Joshua Creighton Allison and Bess Waldon Daniels, and three-year-old sister, Loraine. The story of their last hours illustrates the sorts of agonies people went through. Bess reached a lifeboat with Loraine only to realize Hudson Trevor was missing. She couldn't leave without him and dragged Loraine off the boat with her while father Hudson searched for his son. Neither knew their son was already safe with Alison Cleaver in lifeboat eleven, on the other side of the ship. By the time the parents were reassured their son was safe, their own lifeboat and all others had cast off. So, Hudson Trevor Allison was orphaned, only for tragedy to strike him again. He died aged eighteen, from seafood poisoning, on a beach in Maine. All this behind the single name Hudson in the Allison memorial window.

I had met memories of the *Titanic* tragedy before, when I was part of the team of guides leading Queer history tours of Covent Garden and Soho, London. They included the locations of clandestine clubs such as the Rockingham on Archer Street. The same street has the former offices of the National Orchestra Association, 13–14 Archer St, where the eight members of the *Titanic*'s orchestra rehearsed, on the fourth floor. All perished, and a memorial was placed in the room from funds raised at a concert in the Royal Albert Hall, May 24, 1912.

Why does the *Titanic* story strike us, perhaps more even than September 11, 2001? Because it could and should have been avoided. During the twenty years between remembrance of the orchestra in Soho and the Allison window in Westmount, I discovered more of the story and how it has profound lessons for our time.

RMS *Titanic* was a ship we all think we know something about, thanks to several films like the Oscar-winning *Titanic* (1997) directed by James Cameron. The shock of such a ship sinking is not just loss of life, it is the sense of failure to avoid something as apparently simple as a collision with an iceberg. The liner was made to be luxurious. In fact, aspects of its safety were compromised when more first-class accommodation was built than in the original design. This broke up the segmentation of watertight bulwarks intended to allow the ship to float even if the hull was pierced in parts. So, the rich elite of both sides of the Atlantic went from luxury to death in a couple of hours, and with this came so many stories of "What if?" for who survived and who did not. The Allison family's story is particularly traumatic.

H. J. Allison was one of the richest passengers aboard and unlucky. More working-class passengers died, for whom we do not have such stories. Their cabins were lower in the ship along with all they had in life. Some refused to leave the ship. Did they cling to a myth of the *Titanic* being unsinkable? But the biggest questions for me are still about why the collision happened and especially why the captain was exonerated. The posthumous acquittal of Captain Edward J. Smith was based on the judgment that he had acted within what was established procedure. Captain Smith received warnings that there was ice ahead and behind, from the other ships making the same Atlantic crossing, so he altered course slightly to the south, ensuring lookouts kept a keen eye for ice. This was deemed satisfactory and routine, up to then. Both American and British inquiries found that speed was a conclusive factor. Smith did not slow down the ship due to ice warnings, but there was no evidence the ship was going faster than usual to arrive impressively early in New York (which could have been a temptation as the sea was calm). So, the calamity happened because established procedures were inadequate. It was a collective tragedy.

Since then, a system of iceberg observation and adjusted shipping lanes have meant there has been no recorded incident of a liner hitting an iceberg. Lessons in the specific case of cruise ship safety in the Atlantic have been learned. Sadly, the sinking of the *Costa Concordia* a century later (2012) is proof human error in judging ship safety can still be catastrophic.

The *Titanic* proved massive ships are sinkable and icebergs must be respected. But honoring the tragedy means raising the deep issues that have still not been accepted in practice. The ship represented the high achievements of the Industrial Revolution, to go faster, easier, everywhere: to own and control the earth, or so it seemed.

In a true sense, we live in a titanic age of choices. Just like Captain Smith, we have received warnings, from scientists and nature lovers, that the very equilibrium of life on earth is breaking down and threatening human civilizations. We face different and more complex choices than Captain Smith, but there are real parallels in the false arguments of doing what has been done before or making small inadequate adjustments, simply because nothing awful enough has happened. Yet.

Once the iceberg was sighted, the very natural reaction of the officer on the bridge was to turn away to port, but this meant the *Titanic* was struck on its starboard side, causing worse damage than a head-on collision. Without slowing down, especially with complex systems like climate equilibrium, solutions can be misleading.

It gets worse: the speed of resource depletion is a much more complex indicator to determine than the movement of a ship towards an iceberg. The impacts of human activities have tremendous inertia. Social historians already talk of culture lag in general, and the "environmental" lag is exaggerated by a lack of consensus on the problems and the solutions.

It is terribly ironic that this tragedy was caused by hitting ice, while our efforts are now focused on preserving ice from melting due to global warming. Warnings to slow down, to change course, apply to far more than global temperatures. What of slowing the growth of human population? Understanding how populations thrive with lower birth rates helps. What of building true costs of resource extraction into prices on the market? We will learn how this is possible. The warnings are clear and solutions offered, yet politically downgraded, making this a folly far worse than the loss of the *Titanic*.

The models of how the multifactorial systems of planet Earth are affected by human activity are still so new we do not trust them. The science is at its limits, and because of this it is vulnerable to reckless criticisms and denial. Selfish interests, in defending profit margins and stock market returns, stifle the sort of hopeful and informed leadership we need. Even when a leader is convinced of the degree of the crisis, there is still the "art of the possible" to make change; how to make political change when there is a dependency on carbon fuel, for example. This is a conundrum facing the Canadian federal government and the province of Alberta.

Imagine you are the twelve-year-old daughter of a refinery worker in Alberta, living at Fort McMurray. You have lived through the near loss of your home due to wildfires (especially 2016). Your father earns a good wage, but the lifestyle is expensive, with a mortgage, two cars, a dog, and a younger brother. Your mom stays home until he goes to school. Everyone knows that the margins for profits on oil sands are limited: the industry is subsidized and under attack. There is the sense that it will never be as good as it has been, not unless a pipeline can increase total production. Whom do you think your father will vote for? And if a different party has power, one that prevents the pipeline, so that your father loses his job and you have to leave the province, say goodbye to all your friends, and restart life in Toronto, how will you view "Green" politics then? The human costs to rapid social change, even when it may be for the best, are not to be overlooked. Knowing this is part of the political inertia that is periling the world.

The east side of the sanctuary of Westmount Park United Church is dedicated to remembrance of the two world wars of the twentieth century.

Beneath a large peace window of Leonardo da Vinci's *Last Supper*, with religious and military motifs, are plaques with the names of those who died or served. One family was doubly struck to lose both sons in the last year of the Second World War. Year by year we remember them and are glad to welcome the Westmount Beavers, Cubs, and Scouts for the occasion. Their youth takes us into the future. We also include a video about nature, as we do every week. It is a moving combination. On our lips, as we say, "We shall remember them," is also the pledge "never again." The Great War, World War I, was the war to end all wars, yet its "peace" was false, and it created the conditions for the rise of Nazism.

The *Titanic* tragedy is an illustration that just being industrial is hopeless—ill informed. Green includes the drama of huge social-political change, to integrate our knowledge; it is the human arena of a revolution as significant as the agrarian and Industrial Revolutions before it. Only the time frame is shorter. The agrarian revolution took place over three hundred to five hundred years, the Industrial Revolution two to three hundred years. For a "Green" revolution: less than a century.

There is a spirituality to this, with new values and new consciousness, and no one has a monopoly, especially the church. We already see and understand more, thanks to landmark technical changes: space travel gives us an image of the beautiful blue planet that is Earth, and the Internet and World Wide Web and the intelligent mobile phone have so altered individual human experience. We answer "Who am I?" and "What is the meaning of life?" very differently than our grandparents! Probably with more freedom. Now we must let our answers inform our behaviors and systems.

To honor the *Titanic*, we can quit the trust that "business as usual" is unsinkable. We can respect new knowledge and talk of our destiny to make the planet "home for all."

How we get there is the gospel of Jesus Green.

NEW CONSCIOUSNESS, NEW GROWTH

I am glad to be born at time and place where the suggestion to hail Jesus Green does not risk my own life, nor expulsion from the church I serve. When you give a new name to something that is well known, it is not a small matter. It may be art, in a cool postmodern fashion, and we know naming changes perceptions. But for Jesus Christ this is a huge statement. It is an invitation to be open to new awareness at this critical time of human history. It suggests the sort of change of mindset that is key to new living and new happiness.

With Jesus Green there is a direction to the teachings of Jesus, so what it means to be human may correspond with the story we have of life on earth. Neither Jesus teachings on the realm of God, nor the origins and development of living things are straight lines. In the twenty-first century, both confront us with warnings that growth must have limits and be based on the interdependency of the living world. This sets new goals and satisfactions, fresh appreciations of wealth, beauty, human genius, and the wondrous world. Although this comes through dire warnings, integrating faith and science is not to go backwards, but to be fuller human beings; richer than ever, happier.

I put it this way because the narrative of growth is usually dominated by those opposing change. They define the terms and present questioning the status quo of economics as costly, threatening to daily wages and meaningful employment. But growth and change are natural, essential even, to human endeavors, so it is wicked to insist growth be measured by material production alone. Terms for growth must diversify, or we set ourselves up to fail. (The UK government is among those making progress.)

You are probably ready for change: Jesus Green is a preposterous title for a book, yet you are reading these pages. What you already know of Jesus may help or hinder my task, but the *church* of Jesus can add difficulties. Some are so "heavenly minded they are no earthly good." Still worse, those who grab headlines for their Christian convictions are associated with racism and white supremacy, or a literalism that shouts of a lack of education and frustrated life conditions. If that is not close to you, how about the churchy reticence to admit the joy and passion of living, so that a bishop delivering his wedding address on love to a royal couple makes waves.[5] These various profiles for negative views of the church are tough to change. But one you have control over is your expectation that church should be perfect.

A Chinese couple came to me to be married. They had met through work-study experience as immigrants to Montreal and had no Christian background, simply a respect for spirituality and an appreciation of our church building as sacred space. Even the most basic beliefs about Jesus escaped them. I began to fill in the gaps and realized it was too big a task, not in a wedding preparation process. Then I asked myself, suppose I met St. Peter, or Judas even, wouldn't they think I was woefully ignorant? Perhaps they would understand I depended on other people telling me the best they could.

---

5. Parry, "There's Power in Love."

INTRODUCTION

The understanding we have of Jesus is incomplete, sometimes plain wrong, and possibly twisted. God help us! Authentic faith works despite us, as well as through us!

The way I tried to tell this young couple about Jesus was for this century. Right or wrong, I picked out what I felt was essential and helpful, the leading of the Holy Spirit in conventional language. Jesus' story never ceases to surprise me as I retell it. There is a living quality, because it rests on human existence and goes back as far as we can identify human consciousness, with the tokens buried with our prehistoric ancestors. I practice this each Easter, when I give out mementos to worshippers to claim "Christ is risen!" Such as a square of white silk carrying the image of the face of Jesus, à la Turin Shroud. "Jesus lives through time" would be the message. The Turin Shroud was made through the medieval worldview that insisted on a bodily resurrection. There are still millions of Christians who find this traditional metaphysics of heaven above earth helpful, essential even. Others of us realize, like the shroud, this is a construct for faith, not the essential thing. For what is essential in the Jesus story is ourselves, you and I, our time and place and how we let the story affect us. This is how the shroud works, too, evoking faith that Jesus lives, without being able to hang onto what this means, as we surf the wave of the frontiers of time, telling us about quantum theory, life on Mars, and the human subconscious.

For the first Christians, they identified Jesus of Nazareth as the Messiah, or Christ, the anointed one, with all the inherited meaning that their faith claimed for him. They did not have all the answers. They gave Jesus the title Christ because it fitted the hope, courage, and love they had found. Even then, a crucified Messiah was out of any Jewish expectation. It was appallingly new. But new life and new thinking rose with the crucified Jesus and the people who followed his way. This naming was part of the new consciousness that Jesus gave to his followers.

Jesus Green is a name that may emerge as we travel together through the insights of what we call Green and realize Green itself is growing and changing as our global consciousness reaches new intensities. This relies on there being a true resonance with Jesus because he had sown seeds for the new consciousness of being human that comes with Green awareness. Take this story from the Gospel of Matthew, Jesus' parable of the workers in the vineyard:

> For the kingdom of heaven is like a landowner who went out early in the morning to hire laborers for his vineyard. After agreeing with the laborers for the usual daily wage, he sent them into his vineyard. When he went out about nine o'clock, he saw

others standing idle in the marketplace; and he said to them, "You also go into the vineyard, and I will pay you whatever is right." So they went. When he went out again about noon and about three o'clock, he did the same. And about five o'clock he went out and found others standing around; and he said to them, "Why are you standing here idle all day?" They said to him, "Because no one has hired us." He said to them, "You also go into the vineyard." When evening came, the owner of the vineyard said to his manager, "Call the laborers and give them their pay, beginning with the last and then going to the first." When those hired about five o'clock came, each of them received the usual daily wage. Now when the first came, they thought they would receive more; but each of them also received the usual daily wage. And when they received it, they grumbled against the landowner, saying, "These last worked only one hour, and you have made them equal to us who have borne the burden of the day and the scorching heat." But he replied to one of them, "Friend, I am doing you no wrong; did you not agree with me for the usual daily wage? Take what belongs to you and go; I choose to give to this last the same as I give to you. Am I not allowed to do what I choose with what belongs to me? Or are you envious because I am generous?" So the last will be first, and the first will be last. (Matt 20:1–16)

This is not the twenty-first century with fair wages. The daily wage was subsistence pay, like daily bread. If you make the wage, your family eats; otherwise they go hungry. Imagine the joy of the workers who find even arriving late in the day gives them enough to live on. Now with contemporary mindset imagine the happiness of those early workers who are able to accept the generosity of the owner rather than holding on to their grumbling; their happiness comes from enough being enough; there is enough for all and an escape from an anxiety of subsistence. It fits the slogan you can put on a fridge magnet: "There is enough for everybody's needs and not anybody's greed" (attributed to Mahatma Gandhi). The story assumes that the vineyard is fruitful, with good water, good soil, stable climate, and it places basic human dignity as an integral feature of God's way.

Are you greedy? I see how we are influenced to define happiness through material things, and we live a lie of limitless resources. The environmental movement and fight for eradication of poverty are wedded together in this story. We can be generous, and we can accept enough is enough. It provokes us to place ourselves within it, understand the different reactions, and then act in our own contexts for Jesus Green.

If Jesus told this parable, he had had no idea it would be taken in the twenty-first century to address a global crisis of human sustainability. Yet, he was teaching at a profound level of human nature, which crosses generations, and that is the point.

We are going to consider the truth of our human condition and our place in life on earth with the word "Green." We can name Jesus Green because our human nature is growing in this crisis, to discover a new consciousness, a new joy in being human, being alive. Daring to claim Jesus Green is to exercise a new consciousness that, in our own terms, matches the joy and praise of the first Christians.

I began this introduction admitting this is a wondrous world, despite its sadness and sufferings. It is wondrous to admit the openness of faith and move on from a literalist, pre-Einstein, prescientific spirituality, to discover ecosystems of energies in myriad forms of matter and spirit. It is wondrous to experience how these are codependent and cocreative. This is a Greening of faith into terms of courage and change.

My first thought of "Oh, Jesus Green, that's an interesting and provocative title" has deepened. Rather, classically, I am gripped by Jesus Green, and hope you will be too.

Chapter 1

Which Jesus?

SOMETHING PAST, SOMETHING NEW

Have you stepped through a doorway and—"*Surprise!*" You are framed, on stage before friends and family. Head spinning, you strain to take in twenty faces at once. Happy birthday! It's party time. Were you ready? Astonished or appalled? Your friends who made this happen took a risk on your reaction. Why bother?

But birth is special. Pinch yourself: life is the biggest surprise of all—all routines to be questioned in its name.

Like it or not, each of us has a rapport with Jesus because of birth, and as soon as I put it like this, perhaps you are more open to consider that Jesus could have many versions, at least more than you were taught as a child.

If you are Christian, surprise! This book isn't like the rest. Jesus Green may put your beliefs on stage, in the privacy of your own mind.

Surprise! Jesus Green is for the curious as much as the Christian. We face our mortality, the mystery of birth and certainty of death, with celebrations, contradictions, and coups de coeur.

This personal cocktail of humanity comes together every year with the birthday party for Jesus we call Christmas. The midnight Mass, "of all the parties, in all the world. . ." is a pinnacle of mixed motives and family dynamics. It is a time the congregation can resemble a sitcom. I love it: strangers uncertain to find their seats, talking loudly and dropping hymnbooks, the long-distance Christians in church once a year, and thoughtless "worship-crashing" drunks from the pub next door. All for baby Jesus, the Jesus who is safe enough to bring families together. Some say it is a pity he had to grow up.

Jesus Green is all about stories with many birthdays, weaving your story with mine and Jesus whose story has never been in his control. It belongs to the big story of life, the universe, and everything.

Yet I cannot do this without getting down to details. Humanity keeps pushing for more details, from quantum steps to the infinity of a big bang. To say "happy birthday" truly, to understand someone, or to be at home with myself honestly, details count and mark out my being alive with exquisite truth. The songs that catch this take my heart, and the paintings mark my memory, films too! If I watch the Oscars ceremony I cringe at the schmaltz, yet there is always a moment, despite pretense and popularism, when someone's genius touches me: they capture the ordinary as special. Because it is. That joy, the overwhelming, heart-lifting expansion of being that is life breaks out again. This Hollywood birthday stage is a rarified human product that makes big surprises come true and brims with the paradox: the Hollywood endings tell us life is not like Hollywood. Have you connected with someone, on-screen, because they captured some profound authenticity?

What if Jesus is this?

Boom!

I have opened the door, to surprise you. Has it gone wrong, like an ill-timed joke? Did you expect this all along, as a clever attempt to make you drop your guard?

Personally, I don't like surprises unless care is part of the surprise. If I want to surprise you with Jesus Green, I mean it with care: to respect your story and how you choose to weave it with Jesus Green. It is something past and something new that fulfills a deep yearning for creativity and hope. Let that sentence sit with you a while.

Surprise: Jesus had pubic hair. He went through puberty, not just sexually, but in how his innocence was broken by violence and poverty. He learned how desire had become warped and horrific in human history. He learned to speak of right living through paradox. He said being human demanded more than people believed possible. At least that is the impression I have of him.

But what do you or I really know about Jesus? I hope you will be happily surprised, and when an "easy-access" Jesus disappears in the details, trust me I care.

MANY MAKING ONE

Before you opened this book or considered the title, *Jesus Green*, you had met many Jesuses. Not just different people named Jesus, nor the thousands who take his name to cope with their mental illness. The church gave us a multifaceted Jesus and looked the other way.

It is not just Jesus who multiplied. British king Henry VIII caused such upheavals in history that stories of him are retold every few years. New aspects come through the eyes of someone so far overlooked and the tension to keep to the facts. The TV series *The Tudors* (2007–10) took the troublesome desire of Henry to present a sexy, vibrant king.[1] The viewing figures confirmed it well judged our interests, but it sat lightly to history and became less credible as Henry grew older, with layers of makeup and padding. Few of us cared by then; we were into the story again.

With the book and TV series *Wolf Hall*, we are focused on the life of Henry's eventual chancellor Thomas Cromwell, with the pressures and sheer terror Henry's power could provoke.[2] It's more historical, serious, and even disturbing as part of the history of British life. I turn my head away at scenes of executions. I am not alone: on May 19, 1536, when Ann Boleyn was executed at Tower Green, Henry sat inside at the back of the chapel of the White Towers.

Even though I touched the very doorways Henry VIII touched and I found myself in the world of Wolf Hall, with a visit to the rooms of Stephen Gardiner (Thomas Cromwell's archenemy) at Trinity Hall Cambridge, stories of what actually happened, why, and how remain in the hands of a few people who wrote them down. They show the personality of Henry VIII was complex and differs according to who told the story. Historians would wish to keep their heads! Then, the Henry who rests in my head is a person that reflects something of myself, not just the sources. I want to like him a little, so I think of him as changed by a horsing accident, to understand his brutal attitude to his wives.

To parallel Henry with Jesus has its limits. Henry matters less. I never lost sleep over these contradictory accounts. I was not brought up to say prayers to Henry, and his stories were never treated as sacred texts beyond criticism.

Surprise! Dinosaurs helped to open the pages of the Bible. The study of fossils led to critical rereadings of the biblical creation stories with questions of how everything began. In the early nineteenth century, a revised dating of the earliest life on earth went way beyond church time lines. The seven-day creation myth was increasingly contradicted by evidence that life evolved. New freedom to study the Hebrew texts showed God has two names in Genesis, one for each creation story. The first five books of the Bible draw on

---

1. Hirst, *Tudors*.
2. Straughan, *Wolf Hall*.

four sources, not one. The evidence for the processes behind the production of Scriptures helped their interpretation. We found richer meanings and lost monolithic truth.

Many of the principles established by Old Testament studies were transferable to the New, but scholars hesitated. Who were they to question church teachings? Even now, ask yourself how you read the New Testament. If you are a Christian, it predisposes you to give some of the verses a kind of protection from scrutiny that they arguably lack. We are still debating the miraculous conception of Jesus to a young woman ("virgin" in Greek) named Mary. The author of this story, named uncertainly as Matthew, wrote about Jesus as a new Moses figure. To do this Matthew picks out features of Moses in the life of Jesus. Neither knew their real father. Both were rescued from slaughter as babies. Jesus gave a new law in his Sermon on the Mount as Moses gave the commandments to the people from his experience on Mount Sinai. The miracle of the virgin birth fits into Matthew's particular slant on "Who was Jesus?" We are not dealing with a simple historical account when we read a Gospel but entering sacred history. The pen of the writer seeks you out, with extraordinary claims.[3]

Why does the church, Orthodox, Roman Catholic, and Evangelical, still teach the doctrine of the virgin birth to millions of followers? Attitudes towards Holy Scripture predispose the nativity story to be literally true, even though an immaculate conception contradicts known science. It is even a plus: that Jesus was Jesus because of this breaking of the laws of nature.

There are at least two problems here: miraculous thinking was far more consistent with magical and prescientific consciousness and not such a huge demand to the original hearers of the Gospels as for us today. Second, it suggests that God is not at work in ordinary life, pointing to a deity who deigns to interfere occasionally. Is being a Christian a sort of spiritual lottery plan? Belief is reduced to trusting the impossible and outlandish.

Instead, the freedom to be critical uncovers inconsistencies and contradictions between the four Gospels of the New Testament. This makes Jesus a more uncertain figure, but the same freedom confirms other Gospels from Thomas, Peter, Judas, Mary, and Barnabas were excluded, and these made far more miraculous claims. The Jesus we know was not the most miraculous.

The pressure for some decisions on the Gospels of Jesus mounted when Christianity was chosen as a religion of the Roman Empire. There would have to be greater agreement and order through a series of global or ecumenical councils of the church. You may know the first Council of

---

3. Vermes, *Nativity*.

Nicaea of 325 CE because of the Nicene Creed. As these councils created the canons of Scripture they repeatedly insisted on the humanity of Jesus; he was not a divine being descended in human form, like Greek gods. The four familiar Gospels of Matthew, Mark, Luke, and John were selected due to their antiquity and claimed "apostolic" authority.

The simultaneous search for truth and order made these councils determinative for the church, and people have dedicated their whole careers to study them. You will not want to plunge into the complexities, but something like their dynamics is caught in the film *Name of the Rose*, based on the novel of historian Umberto Eco.[4] This medieval murder mystery, with Sean Connery and Christian Slater, is set in Italy in 1327, and revolves around a council called to resolve disputes between Pope John XXII and the Franciscan movement. This was much later than Nicaea, but that life-and-death issues were at stake and that winning arguments did not necessarily produce results is also true of the first ecumenical councils. Creeds came out of hard bargaining.

What I find helpful in a personal reaction to Jesus today was the acceptance by the church of four Gospels, not one. They faced the question "Which Jesus?" and refused to answer it exclusively.

No single story of the life of Jesus was approved above the others. A blended Gospel, called the Diatessaron, by Tatian, was popular, but it was resisted in the end.[5] This is rich. The politics of different regions combined with the understandable reaction of groups to prefer *their* Gospel. The church endorsed the activity of God in history, that just as Jesus was a real historical person, so too were the church communities that grew in response to him. Memories handed down orally and by circulation of separate Gospels or fragments of them were respected, despite some clear contradictions.

The early church accepted the historical Jesus was at a distance, even for them: none of the stories were so complete to be the definitive account; each declares in a unique way that Jesus of Nazareth was the Son of God (Matt 16:13,16; Mark 1:1; Luke 22:29: John 3:17).

All this was of great concern to Victorian scholars as soon as they had the courage to set out in critical study of the New Testament. There was excitement that rational analysis, free from church control, would finally reveal the historical Jesus. Minister turned missionary doctor Albert Schweitzer wrote a notable summary.[6] This describes the work of several generations of scholars, such as David Strauss and Bruno Bauer, to get to the

---

4. Annaud, *Name of the Rose*.
5. MacCulloch, *Christianity*, 181.
6 Schweitzer, *Quest of Historical Jesus*.

man behind the Gospels. Schweitzer gave a convincing conclusion this task was impossible. Nearly ninety years later, Robert W. Funk, Roy W. Hoover, and the American academics of the Jesus Seminar gave us a cautious, qualified "counter sequel" with *The Five Gospels: The Search for the Authentic Words of Jesus*. Eighty-two percent of the verses of the New Testament attributed to Jesus are "rejected," leaving us with only a few pages, and none are guaranteed![7] Other scholars criticized this well-intended, thorough study as too optimistic. It seems the first councils were right not to press further than Matthew, Mark, Luke, and John and to respect that their sacred history has barriers.

Please do not be alarmed, or too triumphant, if you wanted to read a book that dismisses the historical Jesus. There are still historical truths to the story, but fewer than the church would like. You know the importance of choosing quality over quantity. This radical analysis of the Gospels simply puts us on alert.

I claim a new title for Jesus not because it is trendy but because it is faithful to Jesus' life and teachings seen through the lens of twenty-first-century living. It was always so.

Suppose for a moment that Jesus lived and died in a different part of the world, the Arctic. His face miraculously appears once a year, through ice that must be scraped away until its thinness can reveal the dark depths of the waters beneath. The Gospels of the New Testament are four places where the ice gets thin, but you know you cannot go deeper, or you will destroy "the screen" and lose seeing Jesus again altogether. Each gives a different but related image and story.

Then we look around and notice thin ice where it should not be, with the warming climate. A calamity that has been predicted is taking place in our time, destabilizing life on earth. Jesus of Nazareth's predictions of the end of the world resonate again, based on science.

There are many truths to make one Jesus, including Jesus Green.

JESUS AT YOUR SERVICE

Unfettered Bible study certainly puts a challenge to a believer. How the church created and authorized four Gospels is not well known and can undermine a confidence in what makes Gospels true. But think of the alternatives. If the Gospels are to hold "gospel truth" in the details of what was said and done, accurate to the point of being like four cam recordings, then it asks for miraculous capacities and processes.

---

7. Funk et al., *Five Gospels*, 5.

Get four people to describe the same visit of a spiritual teacher, and you will get partial agreement. Now add a delay of decades with several other versions of what happened secondhand. Then, at the stage of putting written and oral memories together, as Gospel writers did, ask what is the motive for the writer, their angle, and their audience. Jesus did not seem concerned about writing his message down, unlike the prophet Jeremiah and the first Christians, who lived with a sense of an imminent end of everything. As the arrival of the apocalypse became uncertain and fewer first and secondhand witnesses were alive, sayings and descriptions from Jesus' life were written down. I can imagine this was controversial and caused more delays.

Try a different story: think of Santa Claus and yourself when you were five years old, then ten and thirty. Perhaps these ages do not match your milestones, but surely you have changed your understanding of Santa as you grew up. As a child or adolescent, we discover Santa Claus is a legend that parents and guardians sustain to celebrate Christmas and enjoy magical thinking with their children. It is good to wonder and play with generosity.

Jesus' story is not so simple to dismiss as make-believe, because adults go on believing in him despite the magical parts to his story. What confusion! I could become frustrated, angry even, to realize the gospel truth is not a literal truth. If I cannot believe it happened this way, what is the point of what is left?

The classic Christian experience, in twentieth-century Christendom, was missing a beat. We failed to teach children to discover a relationship that lies behind the words rather than the words themselves. This is because the adults were trapped too. I know Sunday schools would regularly talk about the "picture" of Jesus in the Bible, and mainstream churches did not fall into literalism, but too often we have suggested it is easy to get to know Jesus directly. The linguistic and cultural differences and the hand of the author were glossed over, for fear of diluting the claims of the story, diluting the faith. Perhaps this naivety was your "teenage" departure point from the Jesus story and church.

This failure to prepare children and teenagers for being adults in believing is like the way sex and sexuality are so poorly dealt with. Just as we live through sexual development, we go through faith development. The hesitancy to talk about sex between parent and child comes from the adults who are uncertain themselves about sexual feelings and values. Why are we attracted sexually and to whom? What do we do with these feelings, and what makes expressing them most of the time so terrible? Restraint and shame about sex are learned as children observe adults. Yet, maturing sexually is inevitable for most of us. Our bodies change: pubic hair, menstruation, erections, ejaculation, wondrous and magical powers in our bodies surpass the legend of Santa Claus. What of growing up with Jesus? Adults are even less confident to talk with adolescents about faith development than they are of sex. We lack the basic vocabulary of development, and without it, how can it be described? We are, in the main, ignorant it even exists.[8]

Imagine how you pictured Jesus when you were ages five, twelve, and thirty, if you are older than that. Chances are, you do not remember any difference in how you saw Jesus when ages five and twelve, but something radically changes in becoming an adult.

This growing up lets us know Jesus better. You decide, but my impression of Jesus is that he would be at our service. He would want us to know the truth, as we question what we were told about him, in our own terms and times. This means Jesus changes if we follow him, and yet church culture can kill this dynamic with a vacuum of false certainty. We sing familiar songs, the format to worship varies little, and the sermon usually affirms the same beliefs, however skillfully they are given with a contemporary twist. Take a look in the mirror, especially if gravity moves your face around. Jesus is in this mirror, too, made in your mind's eye, never finished.

This is a different Jesus than the one trapped on paper in Holy Scripture, but you cannot just make it up: it still takes engagement with the past.

Jesus the Disturber

One of the best attempts to present Jesus to adult critical questioning is given by New Testament scholar Gerd Theissen.[9] This is a fictional account of a Jew forced to spy on behalf of the Romans. We never get a direct meeting with Jesus, and that's why it is about his shadow. Jesus appears to want other people to speak of him in their own way or simply live out their lives from his teachings, and this is good enough, frustrating though that is. Theissen implies the uncertainties we have about Jesus are not just because our sources are limited but come from Jesus' personality. Jesus was far more concerned that we receive his teachings, his parables, his warnings, than that we puzzle about his identity.

As his fame grew, people wanted to know who Jesus was. "Are you the Christ?" was already a question of his generation. There was one phrase in Jesus' responses that had deliberate ambiguity. He describes himself as the "son of man," but he could also have implied that it had capitals, as "Son of Man." This could mean "one such as me" or "people like us," as well as "yes, I am the Messiah or Christ." You would have to be there, and alert, to make

8. Fowler, *Stages of Faith*.
9. Theissen, *Shadow of the Galilean*.

your mind up. Jesus knew "Son of Man" in the apocalyptic book of Daniel anticipated the arrival of the Messiah (Mark 2:10, 28). He seemed to avoid claims that would be too provocative, immodest, or easy for the listener, just as he did not like simple boxed-in answers, especially about believing. It fits the method of a teacher who tells parables that are open to interpretation.

Nevertheless, nineteenth- and twentieth-century scholars were tantalized that despite the layers of transmission and interpretation, "son of man" was a way Jesus talked about himself. I am easily drawn in, too, and the lure of the Jesus of history will never go away.

Perhaps you think I am already assuming too much. You are more skeptical about any Jesus story. We can analyze the Gospels, but was Jesus a real historical person? Could he have been the creation of a Jewish sect to claim the Messiah had come, so their faith could survive the Roman destruction of the temple? I think a historical Jesus is credible, more credible than being fiction.

We have non-Christian records. Flavius Josephus was a first-century Jewish historian who wrote for Roman authorities, and his extensive writings shed light on the impact of early Christian activities. He writes separately about the deaths of John the Baptist and James the brother of Jesus. His words about Jesus are so brief, they frustrate us as much as they enthrall:

> Now, there was about this time Jesus, a wise man, if it be lawful to call him a man, for he was a doer of wonderful works—a teacher of such men as receive the truth with pleasure. He drew over to him both many of the Jews, and many of the Gentiles. He was [the] Christ; and when Pilate, at the suggestion of the principal men amongst us, had condemned him to the cross, those that loved him at the first did not forsake him, for he appeared to them alive again the third day, as the divine prophets had foretold these and ten thousand other wonderful things concerning him; and the tribe of Christians, so named from him, are not extinct at this day.[10]

I find the limitations to this description give it authenticity. Josephus' words have come to us through mostly Christian hands, showing signs of some touch-ups! The tenth-century Melkite historian Agapius gave us the "Agapian text." Yet non-Christian copies of Josephus' reference to Jesus are similar, which suggests that the original is not greatly altered. Some argue, however, that Josephus wrote a paraphrase of Luke's Emmaus resurrection account,[11] but non-Christian sources cross-referencing the historical

---

10. *Antiquities*, 18.3.3.63–64, in Josephus, *Complete Works*, 576.

11. Maier, "Josephus on Jesus"; Goldberg, "Evaluating Josephus-Jesus Paraphrase Model."

existence of Jesus are nevertheless significant (forgive the pun). They are certainly akin to evidence for other historical people we take for granted.

This independent account, along with the unlikelihood of the four Gospels being written without any historical basis, helps to establish Jesus as a real historical figure. The inconsistencies between the four Gospels also make it less plausible an enlightened Jewish circle made up the drama.

A real historical Jesus who did not proclaim his own divinity may disturb you. It was the impression he left with people, which led *them* to claim him Son of God.

That said, it seems the question "Which Jesus?" was present from the very beginning. Even Peter was confused: Jesus recognized Peter and called him to fish for people, and he named him the rock on which the church would rest. But then Jesus calls Peter "Satan" for having questioned his predictions of suffering and death (Matt 4:18–19; 16:18, 23).

What was Jesus' "mission and ministry" strategy? I laugh with those of you who woke up to your own, well-managed plans not being so good after all. Did Jesus work things out as he went along? The Gospels suggest he had a plan that no one understood. Why didn't Jesus leave a book? That would have made things so much easier; a reprintable mature summary of his wisdom, written even in his last days in prison, please! This would have been reasonable. Jesus could have done better. Not to leave written teachings adds to his lack of judgment in whom he chose to be his disciples and his hope, till the last, that his Father would save him from such agony as crucifixion. Such contradictions encourage me to place more confidence in the Gospels as faithful history. You would not wish to portray your hero in a dim light at all. They present faithful meaning out of historical events.

Fortunately, Jesus is not limited to history and never has been. Think of the description of someone being "larger than life." This is an extrovert who makes an impact on those around them. The Gospels are records that seek to introduce you to the "larger-than-life" Jesus while describing a man of prayer. He is an introvert compelled to be extrovert. They tell of a holy drama that disturbed the peace, when that peace was false. While Roman and Jewish authorities struggled to maintain social and spiritual order, people suffered injustice, sickness, and neglect. Life lacked moral sense too. Jesus taught as a vibrant disturber: he disturbed the norms, and the depth of these teachings mean they cross time.

New Jesuses emerge for us:

The "Pro-Women" Jesus

The Gospels tell us Jesus made important friendships with women, which continued through the church. The bravest of them drew on this female-positive Jesus to affirm their own status, like Hildegard of Bingen (1098–1179), or to support their descriptions of a maternal God, like Julian of Norwich (1343–c. 1416), so that in modern times a strong feminist Christian movement developed, finding Jesus *Christa* liberating and inspiring. The roots of this *are* arguably from Jesus. The history of the church shows it lost Jesus' countercultural insights and became littered with abuses towards women, even as institutionalized persecutions in the period of the Inquisition with the hunting down of witches.[12] You can also point to verses in the New Testament that counter Jesus' welcome. A give-away that people did not fully appreciate or accept what he brought.

Paul wrote women are to be silent in church (1 Cor 14:34). Why order this unless they were already speaking? I am not going to offer proof that this sort of debate is over: I agree that you can find a New Testament verse to justify your views on men and women in the church, whatever your position.[13]

It is the typical, categorical, and unequal thinking about male and female roles and responsibilities that makes the stories about Jesus' attitudes and actions towards women significant. Luke's Gospel has most to offer: Jesus ministers to women in public and relies on their aid (Luke 8:1–3); he takes a woman as the subject for his teachings (Martha and Mary, Luke 10:38–42), and women are the first to realize an empty tomb (Luke 24:1–10; Mark 16:1–8). The Gospel of John remembers Jesus met a woman at a well (John 4:5–42), and Mary Magdalene is the first to meet the risen Christ (John 20:11–18).

Then the letters, or epistles, sent to the first churches before the Gospels were written, show women had leadership and responsibility. Women like Priscilla, Claudia, Euodia, Syntyche, and Apphia are named in greetings and farewells. Phoebe and Junia are given special mention as colleagues by Paul, in his letter to the church in Rome (Rom 16). These are people in their own right, not named through reference to a man.

Anthropologist Bruce Malina helped me to realize the culture shock I would have to go back in time to Jesus. People knew themselves by how they fitted in with others, in a series of different roles. So maybe I would be lonely. Malina sums this up as dyadic or pair-based personalities. Parent-child,

---

12. Hill and Cheadle, *Bible Tells Me So*, 35.
13. Dinkler, "Bible and Women?"

doctor-patient, individual teacher-pupil relationships are still part of our lives, but in Jesus' day the idea of there being individuality apart from these was strange. That made public life more important. A contemporary parallel is the growing angst of online social media. If you think cross-dressing is provocative, even crossing the street was noticed then because everyone was affected by what others were doing, all the time! The struggle to maintain identity and social status, in terms of honor, made this an agonistic culture. While we work to undermine prejudice and racism, then it was encouraged. A person would know themselves by belonging to sets of groups. Hence the Gospels refer to the Samaritans, the poor, the Pharisees, the tax collectors, and prostitutes: categorical thinking was normal. Male and female roles were key to this matrix, which makes the pro-women stories in the Gospels truly remarkable, and a phenomenon around Jesus' life and teachings.[14]

In the difficult process of any attempt to describe what Jesus was like, the low-hanging fruit are the countercultural features. They make the case for a feminist Jesus, given the Christian practice of exclusion and misogyny over the centuries that followed. Jesus was exceptional.[15]

The Jesus Biased to the Poor

In a similar way, those reading the Gospels from the perspective of the Industrial Revolution and the rise of capitalism found Jesus' words on social justice, money, and the poor inspirational and directive. As with women, Jesus left teachings and actions that cut across the attitudes of his day. Take the assumption material blessings reflected God's. Despite the book of Job, this was common and easy to justify from Hebrew Scriptures (Gen 25:8; Exod 20:12; Deut 5:33; 34:7; Ps 91:16; Prov 10:27). Yet Jesus made a fool of this assumption. The image of a camel passing through the eye of a needle is brilliantly impossible. Who but Jesus would have made this up, to attack a presumption to piety for the top dogs of society? (Matt 19:24). As with Jesus' conversation with a Samaritan woman in public, this exchange with a rich young man left his disciples astonished. The Gospel writers record the details and the reactions.

It was understandable therefore that widening gaps between rich and poor in the late nineteenth century would provoke some Christians to insist on social justice. We have a sense of this with Charles Dickens and Victor Hugo, but it was organized, and named, as the social gospel movement.

---

14. Malina, *New Testament World*, 62.

15. Fiorenza, *In Memory of Her*; Ruether, *New Woman, New Earth*; Webster, *Found Wanting*.

Again, in response to absurdly unjust systems and human suffering, the theology of liberation was developed with Latino theologians Gustavo Gutiérrez and Leonardo Boff in the era of the Nixon and Regan presidencies. This points out the material and social benefits of healings along with Jesus' teachings of justice: the lame, leprous, or blind would know material restoration, along with their reintegration to society. Even a Church of England Anglican bishop joined in when Margaret Thatcher's monetarism was causing such hardships in Britain.[16]

Jesus always disturbs for the cause of the poor and liberation theologians know it.[17] His beatitude "Blessed are you who are poor" (Luke 6:20) has a spiritual parallel, "Blessed are the poor in spirit" (Matt 5:3). But which do you think was original? Do you think praying for daily bread would limit blessing to those poor in spirit rather than poor in goods (Luke 11:1–3)? In the idealized account of the early church, the Acts of the Apostles (from the author of Luke's Gospel), the attraction of the first Christians included a generosity to share wealth as well as prayer and praise (Acts 2:44–45). For Luke, at least, Jesus would have no one suffer in poverty.

The Jesus Biased to Same-Sex Love

I once created a public banner that caused waves in Montreal: "If Jesus loved John, then why not Adam and Steve, Mary and Eve?" It came with the woodcut image of Adam and Eve, by Albrecht Dürer, doubled up and rearranged as two same-sex couples. The nudity was attacked, but it was an affirmation of same-sex love that offended.

Homophobia is still practiced by some Christians. The scriptural references taken to justify it are few and rely on legalism in the Old Testament and particular translations in the New. Often debate will force conservative arguments to rely on Genesis as a definitive statement for what is, or is not, "natural." Nothing in the words of Jesus finds same-sex love at fault. Rather, they encourage a new reading in our terms, for today. That Jesus loved John; that Jesus was unashamed of gossip to heal the Roman centurion's lover, a highly prized slave (Luke 7:1–10); that his command to love one another is an inclusive command, to approve authentic partnerships and friendships, regardless of genders of all kinds. Apart from religion, there is a resonance that non-Christians can grasp far more readily: how the religion of Jesus'

---

16. Sheppard, *Bias to the Poor*.
17. Gutiérrez, *Theology of Liberation*; Mosala and Tlhagale, *Unquestionable Right*.

day excluded those in need, despite common sense, and how the same practice is often repeated by his followers today.[18]

Underneath these specific priorities, for women, social justice, and same-sex love, this reading of Scripture reveals how Jesus calls Christians to be in the vanguard of social change. He challenged attitudes and actions. Jesus taught against discrimination and prejudice when this was the normal convention, in the juxtaposition of Jews versus Samaritans, rich versus poor, well versus sick, pure versus impure. His service to humanity continued through his resurrection: the risen Jesus still bore the disabilities of his crucifixion. God did not seek the body beautiful with the resurrected personification of God's nature. This has an affirmation to everyone who lives with disabilities themselves.[19]

The original Jesus addressed the human dilemmas that persist over times and cultures, as surely as the sun rises each day. The evidence points to him encouraging the biodiversity of humanity. Jesus disturbs us, as he did the first followers, if we are willing to put ourselves in tension, reach to the past, confront the present, and accept the future as open. This is the Jesus who trusted his teachings to imperfect followers: this service as an utter statement of belief.

CHAMELEON JESUS

How chameleons are able to change their color is remarkable: the pigment within their cells moves, and they can change the reflection of internal crystals, too, all in an instance. It appears truly magical, and this is more to be seen, to communicate, than to blend into the jungle. We can find Jesus does both.

The Gospels never give us a visual description of Jesus. His appearance rests in the eye of the beholder. Given the Western origins of many churches around the world, often his statues have blue eyes. The blue-eyed Robert Powell gave the archetypal performance of Jesus for my generation, in the TV film *Jesus of Nazareth*.[20] Director and actor apparently agreed this Jesus would never blink on screen, which added to the power of his gaze by a blue anachronism that seemed to justify all those statues. If Jesus said he is found in the lost and the least of humanity, it should downgrade our search for his image (Matt 25:40). It has ever been so. Florentine painters put Jesus

---

18. Boswell, *Christianity, Social Tolerance, Homosexuality*; Goss, *Jesus Acted Up*; Alison, *Faith beyond Resentment*.

19. Eiesland, *Disabled God*.

20. Zeffirelli, *Jesus of Nazareth*.

in Italian modes. I think the Gospel writers would be happy too. They want us to drop our filters, to receive the teachings of Jesus in the present tense. So, take Jesus to your context and discover how he changes in response to what is going on.

Christians have discovered when Jesus blends in, he also "speaks," just as the chameleon changes to communicate.

Blending Jesus, the editing of Gospel stories to the context of the hearers, has been assumed and analyzed. Take the explanation of Jewish kitchen customs in the Gospels (Mark 3:7–9). Mark knows his community includes many who are not Jews and who need commentary to understand. My favorite is in code: the story of Jesus healing a demon-possessed man. The demons' name is Legion, easily suggesting Roman legions are demonic. They are cast out and allowed to enter pigs. Augustine of Hippo (354–430 CE) took this literally, and sadly he taught this showed animals had no status before God. But notice how the story bristles, once you know the mascot of the tenth legion, posted in the region, was a pig: the presence of Legion within the pigs, animals Jews saw as unclean, was unbearable. The pigs preferred to drown (Mark 5:10–20). This recalls Pharaoh's armies drowning when the Hebrews escaped from slavery in Egypt. The coded context even affects how you read the whole Gospel. At the end the disciples are told to go back to Galilee, where the story started: "There you will see him" (Mark 16:7). This also points the reader to go back, now to hear Mark's message knowing Jesus as the Son of God and Messiah, because of the crucifixion (Mark 15:39). The power of God is relocated to Galilee.[21]

Given the Romans' wish to stamp out any rebellions, before and after the destruction of Jerusalem in 70 CE, we can appreciate the situation made Mark use code. He did not want his Gospel to be a death warrant for those carrying it. Decoded, it reads as a nonviolent resistance Gospel for a new era of God's realm, freed of Roman and temple domination.

A very different context of Jesus in the twenty-first century also has global apocalyptic potential but not by assuming the metaphysics of Jesus of Nazareth. The Jesus who can call the ordinary, the excluded, the nonreligious, and the dispossessed calls from the ground up, from the rising of the sun—now, not then—from the inner resonance of what it is to be human and dependent on the rest of the living world.

A chameleon Jesus is not a weak Jesus. A search for the original Jesus is a sort of divine treasure chase, at the end of a mythic rainbow.

Instead, I propose we welcome the crowd of storytellers, translators, and audiences, with some basic assumptions; Jesus of Nazareth understood

---

21. Myers, *Binding the Strong Man*, 399–401.

the essentials of what it is to be human and taught it in ways that got attention. This explains how there is still an authority to what Jesus is reported to have said even when it is usually not his words. Gospel truth is carried in the meanings, not the literal words.

The renowned theologian Paul Tillich (1886–1965) clarified this in his existentially based theology: God is the ground of our being. The Son of God is likely to address us, not through the ever-changing cultural and religious norms but through the irreducible facts of being alive, being human. His book on faith is on being, rather than on believing.[22] This is the thread to guide us as we lose an easy literal confidence in Jesus, to find Jesus Green. Our challenge is to discern this chameleon Jesus and embody his Way to the full.

Shakespeare asks, "To be, or not to be?" But what about "to sing"? Do you enjoy karaoke or follow talent shows like *The Voice*? We let singers into our intimate selves. Song fills my home. My partner sings better than I, but our birds do it best. Often our mornings begin with the pure joy of canary song, flavored by Mozambique finch "backing." Canaries are extraordinary birds. They molt often, and extremely, replacing almost all their feathers. Singing stops while they molt, as in the wild they would not escape predators so easily. The recovery of their song is not just memory; the song can change when brain activity reconnects with vocal muscles. Canaries sing with a vitality that is real. This is part of why song is so important for us. In song, we can bypass the restraint of intellect and subconscious parental voices with the truth of our "inner child." We restore our animal selves. To make a song our own song, we bring our interior life to exterior reality. This experience of depth can be shared and is what makes song so powerful. Louis Armstrong sings, "What a wonderful world . . ."

As a full-blooded British Methodist, I was born in song, age twelve, praising Jesus in the Royal Albert Hall. It was pure gift for me, like "blessed are the pure in heart, for they will see God" (Matt 5:8). I still have this expansion of being alive, connected, and in love, when a hymn takes off in any church service.

This is dangerous territory. Hymn singing is something that has turned people off religion. Poor singing, detached from meaning and emotion, is widespread, and hymns are peculiar, not for everyone. Like Shakespeare, but often banal. Why don't we all enjoy Christmas pudding? I do not trip out on Marmite. But we must eat, and sing, in our own ways! Even for the gourmet singer, hymn singing is not a routine high: how the words, the music, the singing, the context or meaning of the whole religious occasion

---

22. Tillich, *Courage to Be*.

come together is like waiting for a set of green lights; go through here, they say, and you can arrive at a familiar place of bliss, giving food and drink to the soul that has no weariness or limitations because of life's difficulties.

Often, when a hymn works it addresses a seemingly overwhelming need or challenge and contradicts a sense of powerlessness or failure. "Take it to the Lord in prayer" through singing the hymn is a felt truth of new perspectives.[23] It is no coincidence that one of the best hymns of Methodist hymn writer Charles Wesley, "Come, O Thou Traveller Unknown," is based on the story of Jacob wrestling with the angel of God (Gen 32:22–32).

Jacob has reached a crucial moment of his life when he decides to return home and meet his brother Esau, not knowing whether he would be met with a welcome or death. Over twenty years have passed since Jacob has cheated Esau of his birthright. On the eve of his meeting with Esau, Jacob has a strange divine-man encounter, without explanation. He wrestles through the night. He asks his opponent, "Please tell me your name," but the wrestler refuses. Jacob gets the blessing he seeks, at the cost of a bad hip. The next day goes well.

Charles Wesley takes this story to his experience of God, as a perfection-seeking Christian, juxtaposing weakness and strength, difficulty and ease. There is truth to sing this hymn with its insight of the relationship between the nature of something and the name it goes by.

As Jacob wrestled, he demanded to know the name of his opponent: to know the nature of this stranger. But it was Jacob who was changed and renamed Israel. He would become the father of the nation. Generations later another divine encounter changed a founding figure: Moses is commanded by God, speaking through a burning bush, to go back to the place he escaped from, to demand Pharaoh to let the descendants of Jacob go. Moses asks, "Who shall I say sent me?" "Tell them 'I am' sent me," says God. The original Hebrew is better, because the verb has more than one sense to it: "I am" is also, "I will be who I will be." There is much in the name (Exod 3:1–14).

Postmodernism agrees with this biblical wisdom of names and nature. In our first months of life, we move from a world of imagination to a world of external realities in which we find ourselves (such as first seeing ourselves in the mirror). This is a lifelong process of self-understanding. Even our subconscious follows signs and symbols; it has a language.

To learn self-differentiation from the world and the makeup of the world around us, requires subject object referencing and the choice of words can facilitate or inhibit this. If I have two or three words for snow, I am likely to notice snow less than someone with fifteen. When I learn new

---

23. Scriven, "What a Friend," stanza 3.

names of trees, I learn how to distinguish those trees and I notice those trees more than before. To name something, or someone, changes my experience. This lies behind what I invite with the title Jesus Green, so that Jesus speaks afresh, and we change too.

In the case of Charles Wesley's hymn, Charles shared his own spiritual development. He and his brother John failed as missionaries in the New World. On one crossing John realized his faith lacked the depth he saw in others who faced the same storm. Back in England it was a time of risk as well as revival. The brothers found spiritual renewal and battled establishment. Jacob's wrestling became Charles' struggles. There were fourteen verses in the original poem. The last line of each takes us on a passionate journey of revelation: "Till I thy name, thy nature know" becomes "Tell me if thy name is Love," then "Thy nature and thy name is Love." This is the Jesus thing, to go further than Jacob, somehow to claim a revelation of the ultimate: the existential addresses the essential.

> Lame as I am, I take the prey,
> Hell, earth, and sin, with ease o'ercome;
> I leap for joy, pursue my way,
> And as a bounding hart fly home,
> Through all eternity to prove
> Thy nature and Thy Name is Love.[24]

You picked up this book. You are curious to discover Jesus Green, but a chameleon Jesus may seem impossible to you. You respond to "Which Jesus?" with "Jesus is the same, yesterday, today, forever." This is an assertion of the irreducible otherness of God, God as "I am." But a wonderful paradox then follows. Out of our subjectivity and mortality, how we relate to "I am!" is this: "I will be who I will be," with many Jesuses to discover in our lifetime. God's answer to Moses in the burning bush was not simply about Godself but about human experience and understandings of God.

Do not imagine, if you could have a "back to the future experience" to meet Jesus of Nazareth, you would be left without doubts and questions. A chameleon Jesus matches the multifaceted nature of his "Father" and encourages deeper understandings of tailor-made gospels, then and now.

I realize at this point you may be totally exasperated. The frustration of being led into an "Alice in Wonderland" world, where words can mean whatever I want them to mean, is not far off.

Yet, the boundaries and rules of reference are there. They are found in lived experience. We find different Jesuses as history unfolds. This is not a

---

24. Charles Wesley, "Come, O Thou Traveller," stanza 12.

sign of weakening faith but an inevitable expression of human history and culture. A "courage to be."

We have broken the prison of thought that religious beliefs are true only if they are fixed. Let me share a way, a literal path, to make this freedom your own.

WALKING INTO JESUS GREEN

One of the famous teachings for an eco-friendly Jesus is Jesus' command to look at the birds of the air (Matt 6:26). Perhaps Jesus had no idea of the raw mortality behind apparently blissful populations, as observed by Thomas Malthus, or perhaps he did and pointed people to something more profound.[25] In Greek, "look at" (*eublemate*) has the sense of "consider," which is to admit you search as you see. We are predisposed to find patterns rather than chaos. Jesus suggested the remarkable life of the birds can speak to our preoccupations: "Look at the birds and don't worry." Natural patterns can speak to our souls.

I invite you to look at Jesus, to rename him Jesus Green, with a reliable way to start:

I have a large freshwater aquarium and delight in the antics of the fish, shrimp, plants, and snails. I let my intuition flow as I appreciate their beauty and think of names. I named one fish, a discus, BG, for Grand Bleu, from the film *The Big Blue*, directed by Luc Besson, shot in the Greek Mediterranean (1988). Another is Ripples, whose red-white markings made me think of rippled ice cream. The largest I named Daedo, after the myth of Daedalus, a multitalented inventor, and a Renaissance spirit before the Renaissance.

I learned about Daedalus and his son Icarus in school Latin lessons. You may know they made a fateful escape from King Minos' palace in Crete by flying like birds. In youthful disregard of his father's warnings, Icarus flew too close to the sun, and the heat melted the wax of his feathered wings. Less well known is how Daedalus had been commanded to solve a lethal problem of the king's sexual excess. The bull-man or minotaur was Minos' son. Daedalus created a maze to keep the minotaur within bounds of the palace basement. Minos then struck an awful bargain with the defeated citizens of Athens as they had killed one of his sons. Each year seven youths and maidens of noble birth were to be sent to be sacrificed until one of the youths was able to kill the minotaur. The victor would marry the king's daughter Ariadne, but she knew to kill the minotaur was only half of the challenge. Unless help was given, the maze would trap the victor as well.

25. Malthus, *Principle of Population*.

When Ariadne fell in love with Prince Theseus of Athens she gave him a cloak to unravel as he entered the maze. The myth reappears in the film *The Name of the Rose*, when Sean Connery portrays the Franciscan sleuth William of Baskerville.[26] He describes Ariadne's thread to his novice monk Adso, played by the nubile Christian Slater, to exit a trickily constructed library. We are entering layer upon layer of stories ourselves!

The fish, Daedo, is the sort of discus known as pigeon blood, but I was wrong to presume pigeon blood looks marbled or tortoise shell. A discus breeder's website explains old Asian slang for pigeon blood is "soy sauce."[27] It describes the freckling of spicing your food!

The irregular appearance suggests the complexity of a maze. All this was going on in my head as I looked at the fish. I saw this pattern only because I was already interested in mazes, the sorts of mazes called labyrinths, that have a real relationship to the Cretan culture of the myth of Daedalus.

Labyrinths and mazes live alongside each other in this history. Yet labyrinths are different. To oversimplify, the difference between a maze and a labyrinth is one of a puzzler versus a seeker. Mazes put choices before the walker, who loses sight of the entrance and exit, whereas labyrinths are open, often flat forms, with a singular, complex path and one entrance or mouth, which is also the exit. The whole form is visible, and a labyrinth walker experiences meditation, not frustration. It does not help that films use maze and labyrinth indiscriminately in their titles, and French uses *le labyrinthe* for both. The fictional American TV series *Oz* (1997–2003) is set in a men's prison, with a gym area and a labyrinth laid over the floor.[28] More often labyrinths are found in churches or sacred natural locations.[29]

The thread I offer to identify Jesus Green is found through following Jesus' teachings and death, with the importance of the home to save us from ourselves. My story with labyrinths is personal and helps to introduce it to your world.

In February 2001, during a three-month sabbatical, I traveled to San Francisco with hopes to write a novel about a clerical spy: a reverend, recruited by the pope to report on the New Age movement. In truth it was a ruse for research to help develop new social and spiritual programs for the LGBTQ+ nonprofit Kairos in Soho, in London, which I directed. Often tourists from California would participate, especially on our Queer history walking tours, and I took up their invitations to visit. San Francisco and Los

---

26. Annaud, *Name of the Rose*.
27. Discus, "Origins of Pigeon-Blood Discus," para. 3.
28. Fontana, *Oz*.
29. See https://www.labyrinthos.net/index.html.

Angeles gave me a fabulous variety of experiences: my first tantric massage from a notable practitioner, Jeff Kramer, sexologist and founder of Body Electric; a proper horoscope reading and meditation to find my spirit guide; and a gay men's spiritual circle that led to a quick trip to Palm Springs (like David Hockney's canvas *Peter Getting Out of Nick's Pool*).

On my last day I visited Grace Anglican Cathedral, San Francisco, to walk their labyrinth. It is the North American hub for a hundred others. I was tired and nonchalant toward what it offered. What more could I take in?

It is very simple to walk a labyrinth, but within a few steps I realized it was more sophisticated than it appears to be. You receive this wisdom quickly as it makes you feel lost and drawn forward at the same time. The path turns you clockwise and then back, 180 degrees. You are taken towards the center and then to the outside: "Have I gone wrong because this is taking far longer than I expected? Should I start over? Do I trust myself?"

I answered these questions, step by step, and eased into the repetitiveness of its circular form. With my last step I entered the center or goal of the labyrinth, gratified I had not been mistaken and fascinated by the easy way I had entered a meditative state. I had ticked another spiritual box. Then the divine joke was on me. As my foot touched the floor of the goal of the labyrinth, the cathedral bell struck. I had been watching my feet for twenty minutes. I looked up and faced the dramatic length of the sanctuary, high altar, and cross. Such a cathedral labyrinth is oriented to point you this way, but my timing was something else. I laughed out loud. "Really? This is too corny." There was no obvious meaning, just a strange coincidence. Crude synergy. Not to be ignored.

The next day I traveled to Montreal to taste its winter and inquire of potential work in the United Church of Canada. The bell was a wake-up call that my visit was significant. I think it made me listen acutely to the conversations and to take seriously the welcome I received. Five months later I had given up my job and started a new life in Montreal. I was aged thirty-nine, not twenty-something. This was for keeps.

Labyrinth ministry followed me to Montreal. I was alert to its potential. My first church was near to Montreal West United Church, which ran a monthly labyrinth ministry. I recommended it to my flock. One member became so enthused, he secretly made a cloth labyrinth at home and presented it to the church. What a gift! In turn, we raised funds for an outside brick labyrinth, to offer walking 24/7. It is still there, despite the church now being a Buddhist temple.[30] When I moved on to be a chaplain at McGill University, I offered the same cloth labyrinth walk, in the student union

---

30. Yuen Kwok Buddhist Temple.

building. It helped me make sense of the contradictions of university life. Different values live side by side, so economics is taught without reference to the limits of the environment: corporate and competitive living is taken as success, while others work hard to get students to think for themselves. Some of the more renegade professors held up the question "What is education for?" So many questions and conflicts. They fit into a labyrinth like a glove.

At my present church we offer walks on the same cloth labyrinth, but this time with the earth in mind. Since 2001, the earth crisis has deepened. The structure of a labyrinth reflects the emergence of life and consciousness itself.[31] Labyrinth walking encouraged the emergence of Jesus Green in me. The rings remind me of a brain that relies on a certain intensity made possible by its folding. This labyrinth is the form I walked in Grace Cathedral, which is a copy of a labyrinth found in the nave of Chartres Cathedral, France (labyrinth construction 1215–21).

Long after I named my fish Daedo, I discovered the goal of the Chartres labyrinth had a plate depicting Theseus and the minotaur. Only the worn stubs of the rivets for the plate remain, but the ancient myth had been integrated with Christian life. It suggests cathedral builders appreciated the folly and tragedy of the story as theirs, too, and in need of reconciliation: limits for limitlessness.

The gospel of Jesus Green is to come home, to arrive at the center of the labyrinth with the earth in mind, to make home for all, not just for humans: a gospel of reconciliation that demands you let go of Jesus as your private spiritual coach and savior.

I dare you to walk a labyrinth with Jesus in mind and admit my question "Which Jesus?" is valid. Jesus is sown into history and slips away because of it. He comes through other peoples' stories, their motives and manipulations.

That we are in a process to value, not deny, the limits that give us life: this is the gospel of Jesus Green. It is time to walk into Green.

---

31. Wright, *Maze and Warrior*; Savard, *Labyrinths and Mazes*.

Chapter 2

What Do You Mean, Green?

IRREDUCIBLE

One of the gifts of youth is a readiness to go with the flow. Aged twenty-four, I returned to Lisbon, Portugal. It was early January 1987, my last twenty-four hours in a youth hostel, when a booming voice filled the room and I met its larger-than-life owner, whom I will call Ricardo. We hit it off, and I returned to London with an invitation to visit him in Brazil, "whenever you want." I was in Lisbon to improve my Portuguese after a memorable two-week vacation that summer and had started language classes in Notting Hill. Ricardo's offer bugged me. Brazil was beyond my budget to visit. Yet this was the largest Portuguese-speaking population in the world. When would I have another chance to go? What better way to learn?

I accepted Ricardo's invitation but fended for myself for the first week of my visit in Sao Paolo, a metropolis encircled by vast shantytowns. Hyperinflation and my rudimentary Portuguese meant it was difficult to enjoy the city, but Ricardo arrived from New York and we headed for his hometown, Cambui, in the landlocked state of Minas Gerais, three hours northeast by road. I was the fourth person in a borrowed van, so sat behind my friends with the luggage. I strained to see the view. Half the journey was by freeway, then we turned off and plunged into a sea of trees, our van as a green submarine. I knew about the Amazon but had no idea how the areas of rolling farmland preserved forests. Ricardo had given me a tape of early Baroque French horn music for my Walkman. The motion of the truck, the surprise of natural beauty after a week of cityscapes, and the joy of the music merged in me with mounting euphoria. Tree after tree, at our speed, was jaw-dropping. This was not a tunnel of trees but a place in itself, the branches playing with the sunshine to reveal the wind—submerged without being wet. The world is too big and too wonderful for one soul to grasp, but experiences like this remind me it is in our blood. One story: to remember, however complex life becomes, the green of plants has made our life possible.

Jesus Green may be an immediate hit as a book title, to pique your curiosity, but as soon as you ask "What is green?," I have trouble answering.

Green is an adjective, a noun, and a verb. You know greenhouse, but what about greeneye (a fish), greenhead (a fly), greenlet (a bird), greenockite (a mineral), and perhaps most obscure: greenery-yallery (from nineteenth-century esthetics). I recall green with a small *g* as a color, unripe fruit, or freshly cut lumber; green for what is full of life and vigor; or fresh and vigorous; something new or recent; also a grassy plain. Are you green with envy or inexperience?

The Green I mean in this gospel is loaded with earth-friendly intentions. But what is the difference between genuine Green or a veneer? The Methodist heritage from John Wesley hands me a tool to resist Greenwashing. Wesley led a reform movement within the Anglican Church that became a Church in its own right. He was familiar with the controversy of opinions and valid questions concerning his own authority. He proposed Christian authority comes from four sources: Holy Scriptures, tradition, reason, and experience. The Wesleyan Quadrilateral acts like a mobile that combines each source to achieve equilibrium. This approach is still regularly used in Methodist debates on social and spiritual issues, to guide the search for a common mind. In a century when adequate global consensus is critical, the model is worth sharing.

The four dimensions that I offer to authorize Green are:

- *Energy*, of biophilia (a love of living things)
- *Movement*, of politics (the environmental movement that includes social justice)
- *Choice*, of work and play, household desires and goals
- *Consciousness*, of dreams and nightmares, with science and technology as a mirror to ourselves and the planet

At the heart of this model, a key fact for each dimension, is that we are animals: we are part of nature, and the new meaning of Green has come from warnings. There is evidence humanity has overshot the carrying capacity of earth. As an archer fires an arrow over her target, each of the different dimensions of Green—energy, movement, choice, and consciousness—tell us we are missing the mark.[1] Yet the growth of humanity is natural as well as

---

1. Dr. William Rees takes overshoot as a fundamental starting point (Rees, "Dresden Nexus Conference 2015," 9:28).

relentless. We are not outside of the history of evolution but are conscious we face a dead end, literally.

Business is at the hard end of the Green revolution: in July 2019, BBC Leicester featured the closure of a fish and chip shop, just two years after opening. The notice on the restaurant door declared: "There's not plenty more fish in the sea." The owners had been on a fishing trip and seen pollution and consumer impacts on fish stocks. Running a restaurant is like being a parent of a toddler who never grows up, so I can imagine the downtime gave them space for reflection, and nature triggered some deep affinities, as well as distress, when they saw how fish ate plastic.

Owner Aatkin Anadakat described his commitment to a plant-based food business as "a nice place to be."[2] You have the sense from the story that there he found pleasure in making the change. Not only could it be good business sense but a better experience of business and life itself.

What difference can one business make? Except there are millions like these, and we are more connected than ever to what trends, and *overshoot* is about *us*, not you or me alone. There is no single answer to what Green means. The different dimensions of Green move us from personal to political, as local and global.

ENERGY AS BIOPHILIA

Energy is the basis of everything. But Green energy is deeply personal. We have it as children: the love of living things. The intensity of life that is found in the newly alive is powerful and, above all, outgoing. There is a thirst to explore that parents of all kinds know needs careful supervision. A seedling deepens roots, explores with tendrils; a young rabbit will taste everything.

The love of living things is *biophilia* in Greek, and refers to the love expressed in friendship, hence the Green organization Friends of the Earth. Biophilia came to modern use with Erich Fromm (1964) and Edward O. Wilson (1986).[3] But you do not have to be a philosopher to realize the basic point. For a child the natural world and human-made environment are not distinguishable but learned, and the delight of discovering other living beings besides mother, besides humans, is self-evident. That joy is part of the magic of children.

It does not end there. Peter Verbeek and Frans B. M. de Wall studied mandrills. These are primates, large monkeys known for the males' colored faces, the wise monkey in Disney's Lion King film. They live in

---

2. Robertson, "Fish and Chip Restaurant," final para.
3. Fromm, *Heart of Man*, 37; Wilson, *Biophilia*.

tropical forests on the East African coast (including South Cameroon and DR Congo) and primarily eat fruit and insects. Like all primates, they are curious, and Verbeek observed a group of juveniles with a mature male who discovered a toad that "unconvincingly played dead." They were transfixed by the scene, and the lucky toad escaped, apparently unhurt. The authors described the energy of the encounter: "If an early sense of *wonder* predicts the good naturalist, these mandrill youngsters seemed fit for a career in field biology."[4] The provocation of a sense of wonder in these and other primates as they explore the living world warrants our respect and understanding. Derision towards a "tree hugger" is shown to be an attack on nature itself.

This biophilia lies behind some of humanity's finest achievements. My grandmother, Clarence Neil, would have enjoyed my travels through Brazil. She did not know Green; she knew greens, thousands of them, as she loved gardening. She loved to see plants grow and to know their Latin names. The Latin was a way for her to recognize the uniqueness of a plant, its needs, and how it would flourish in her care. She loved, and the loving provoked care, and the caring sought out new words. But where does this Latin come from? Scientific study of plants is one of the earlier disciplines that stimulated taxonomy, a discipline of names and types, with Latin as the common language. The love and curiosity for life of early scientists has left us marvelous evidence:

> Carolus Linnaeus: "If a tree dies, plant another in its place."[5]
>
> Charles Darwin: "The plough is one of the most ancient and most valuable of man's inventions; but long before he existed the land was in fact regularly ploughed, and still continues to be thus ploughed by earth-worms."[6]
>
> Isaac Newton, who is famous for gravitational theory by the blessing of a falling apple, is not known for his love of nature, and yet: "I do not know what I may appear to be to the world; but to myself, I seem to have been only like a boy playing on the seashore, and diverting myself now and then in finding a smoother pebble or prettier shell than ordinary, while the great ocean of truth lay all undiscovered before me."[7]

---

4. Verbeek and De Wall, "Primate Relationship with Nature," 18.
5. Fries, *Linnaeus*, 15.
6. Darwin, *Formation of Vegetable Mould*, 314.
7. Brewster, *Memoirs*, 2:407.

Marie Curie: "All my life through, the new sights of Nature made me rejoice like a child."[8]

And Newton's super successor Albert Einstein: "The important thing is not to stop questioning. Curiosity has its own reason for existence. One cannot help but be in awe when he contemplates the mysteries of eternity, of life, of the marvelous structure of reality. It is enough if one tries merely to comprehend a little of this mystery each day. Never lose a holy curiosity.... Don't stop to marvel."[9]

Each acknowledged the child within their work: a pure delight of inquiry, especially in life itself. These famous scientists had a total dedication to their work; it was a passion, to the point of self-sacrifice (literally true for Marie Curie). Science gave them a reason for being that can be described as a vocation, a calling in Christian terminology, that I recognize, having left one (zoology) for another (theology). With this writing I find a middle way. The dynamics of living a "directed" life are certainly energetic.

Pause. Recall the joy of nature. Practice it with houseplants, pets, walks in the park, the flight of butterflies. There is a quality of energy in being earth friendly, which is willing to be bound so that others may be free; the *philia* in biophilia is the energy of friendship. It is a valid test for authenticity and safeguards scientists and the rest of us from selling our souls. We are alert to distortion or manipulation. As the struggle for change intensifies, this self-sacrificial biophilia is vital to retain and celebrate. Career biologists and environmentalists find employment in private companies and government departments, each with its code of ethics and mission statements. Given the hierarchical nature of most organizations it is common that rules and directions change; information can be buried or selectively analyzed, and the scientist is left wondering, what am I doing here? The inner voice, the energy, matters.

Billions of people know about biophilia. The popularity of tele-reality shows of the worlds of vets, pet trainers, wildlife refuges, and zoos reflects a huge affinity between people and other living things. There is a global reach to natural history programs such as BBC Planet Earth (*Great Plains, Ice World, Deserts, Caves*). Even the fearful stories of people being attacked in *River Monsters*, with Jeremy Wade, includes Wade's appreciation of little-known fish and their ecosystems. These TV shows can suggest there's no problem. Biophilia means humanity will have no difficulty in changing, in respecting the limits of Earth's natural resources. But talk with a vet, and you

---

8. Curie, *Pierre Curie*, 162.
9. Miller, "Old Man's Advice."

will soon realize attitudes towards livestock and pets vary tremendously; negligence, carelessness, and cruelty are present, routinely, even towards animals that share space in the household. We should not be surprised. Human brilliance is accompanied by atrocities and contradictions.

This was well put in the sad dynamics of the first human family, according to Genesis. The sons of Adam and Eve fall out because God preferred the best of lambs from Abel, rather than vegetables from Cain; in a fit of jealousy Cain kills Abel and then finds himself pleading before God, "Am I my brother's keeper?" (Gen 4:1–16).

Human rivalry continues to plague any optimistic view of human nature. We may say we love animals and we are Green, but our behavior contradicts us. Our wants and entitlements have priority. Albert Einstein witnessed enough awfulness to come up with another memorable quote: "Fear or stupidity has always been the basis of most human actions."[10]

Biophilia can describe the love we need to get along, especially for human beings to ask new-old questions that qualify this energy, with the other aspects of what I mean by Green: "Am I my butterfly's keeper?" We are aware that destruction of habitats of insects is leading to their decimation: the emblematic monarch butterfly has amazed us with a huge migration pattern and taught about habitat preservation with its specific preference for milkweed leaves as "refueling" points. Monarchs developed this dependency long before Homo sapiens took to ploughing fields. Yet who cares? It is only a certain proportion of us who respond to the orange appeal of monarchs. We are careless about needs of pets and hesitant to boycott Nestlé's bottled water, or baby milk, or PepsiCo, Kellogg's, and Unilever products, for their use of palm oil.[11] There is such a thing as environmental fatigue, or desensitization to alarms, when the consequences are not immediate or in our own backyard.

This disconnect can be more serious; Woody Allen has put words to the experience of biophobia: *"Nature and I are two."*[12] Environmental educationalist David Orr puts biophobia alongside misanthropy (hatred or distrust of the human) and sociopathology (antisocial and lacking any conscience). Yet nature is repeatedly used to sell products. Nature phobia is not as common as the dullness to our surroundings in general. There are plenty of distractions, even when we could connect: the snowmobile culture, mobile phone conversations during a walk in the park, jet-skiing on

---

10. Einstein, *Ultimate Quotable Einstein*, 187.

11. There are thirteen well-known companies in chains of production accused of serious and long-term human rights abuse and environmental damage (https://palmoildetectives.com).

12. Orr, *Earth in Mind*, 131; emphasis added.

the lake . . . The busyness of urban life is so easy to take with us as a bubble of hectic thoughts. We lose touch with basic realities of birth and death, growing and making food, waste disposal and water supply. Nature teaches these things, and with astounding wisdom, if we notice. Philia, friendship, takes time and listening. Where do you place your love?

There is another insidious phenomenon: *environmental generational amnesia*. It profits from lack of attention, parent to child. The ability to notice is taught and gives lifetime memories of joy. But untold stories mean each generation takes as normal the state of the natural world they are born into; the sense of loss that older folk have when they return to the meadows and riverbanks of their childhoods has no impact. Nature's glory slips away, unnoticed.

It is normal to be numb to the glory and pains of the natural world, in the drive to pass exams, get a job, make a career move, win in sports, find the ideal partner, raise a family, and yet, the call of this love is not going away: I can appreciate the drama of tropical forest destruction without seeing it directly. I can lament the bleaching of coral reefs. It gnaws daily at the conscience of most of us, even if we have sensible arguments to ignore it. This is because biophilia is natural, a given trait, available to question every suburban dream.

Raising ourselves to be as Green as we can be is to breathe again, revitalize our humanity, and stretch for progress everyone can smile for. It is the energy to vote for change.

Biophilia is the unrelenting source of Green.

MOVEMENT AND POLITICS

Movement is life. From the control of stomata of plants, to the swelling mouth of a blue whale catching krill, expansion and contraction permit the circulation of energy.

Humanity is on the move. There are eight times more human beings alive today than in 1800.[13] It is the overshoot: a peril to far more than humanity.

As human populations grew so rapidly with the Industrial Revolution, some individuals realized harm was being done to nature. They spoke up for it, made circles of concern, societies, and limited legal reforms.

It is correct to describe Green as a political movement that began with these kinds of initiatives, but there is a continuity with nature that helps to understand the politics. One of my heroes as I found myself both a Christian

---

13. Kaneda and Haub, "How Many People," para. 9.

and a student of zoology was the Roman Catholic priest and paleontologist Pierre Teilhard de Chardin. His *Phenomenon of Man* (*Le phénomène humain*) is an inspiring account of the development of life on earth, because he points out the trends as well as the outcomes. Life is growth in mass, variety, complexity, and interdependence. Life has a direction, not from outside forces but through inherent potentials.

David Sloan Wilson argues there is an evolutionary basis to our behavior with our concerns for the planet and Green politics.[14] I understand this as a form of biofeedback; it reflects the growth and complexification that Teilhard described as the history of life.

All to say, the politics you assume around the word "Green" have a biology to them. To read or hear these words is part of human adaptation! Key events and organizations mark out a history of change and teach us it is possible to make life-changing agreements.[15]

We can begin a chronology of the Green movement in 1642 with forest preservation in Japan. But perception of the human destruction of nature is ancient. Greek and Roman historians describe mudslides and soil erosion around Athens and Rome due to deforestation.[16] Human beings have changed landscapes over millennia, especially through the agrarian revolution with land clearance and changed water courses. The Industrial Revolution scaled up harmful activities and combined it with sudden population growth. We seek solutions for positive feedback to offset this negative one.

Nature welfare and human well-being go together. One of the earliest environmental measures was taken for human welfare in 1863 (UK) with the Alkali Act to reduce the clouds of hydrochloric acid being released in the creation of sodium carbonate. A year later the US Congress assigned sequoia trees of Mariposa Grove in Yosemite Valley for public, resort, and recreational use. In 1883, The Natal Game Preservation Society was formed in South Africa. The National Trust was founded in Britain in 1895.[17]

In 1929 the US passed the Migratory Bird Conservation Act. It took years, but the Soil Conservation Society of America was founded in response to the dust bowl storms of the thirties.

---

14. Sloan Wilson, *This View of Life*.

15. McCormick, *Global Environmental Movement*, 297; Schreurs and Papadakis, *Historical Dictionary*, xix.

16. M. Williams, *Deforesting the Earth*, 71, 74.

17. Schreurs and Papadakis, *Historical Dictionary*, xix–xx.

A historic landmark came with success to recognize and oppose the impacts of the pesticide DDT on birds,[18] then came actions to reduce use of CFC gases that caused the loss of the ozone layer.[19]

The sixties and seventies also saw the first direct action and campaigning groups that transcended national boundaries. The wholesale goals of Green political parties emerged: human societies must be shaped by scientific facts, rather than by commercial "truths." The decisions to combat DDT and CFCs exemplified these goals: given the wealth of history and complexity that comes with them, they are helpful foundations. They highlight the importance of freedom in research and the press to offset pressures from government and commerce. Science, after all, is rooted in the reality of nature.

The role of technology through the twentieth century has become a blessing and curse: the multinational global market has reached into virgin forests and ocean depths, but satellites and Apollo moon projects have shown us planet Earth: rare, beautiful, and shared. Now I can see my neighbors' backyards, all over the earth. The bigger picture of human endeavors is important, with clashes of empires, ideologies, nation-states, the rise of unions and votes for women, two world wars, the Cold War, and Chinese Revolution.

Arguably, change has always been difficult. We have failed to let information affect our systems; important reports like *The Limits to Growth* commissioned by the Club of Rome were resisted, and the Green movement has always played catch-up.[20] This failure to adapt is already making millions of environmental refugees. It now threatens the stability of democratic countries that hold the global village together.

But I get ahead of myself. What also changed radically with the arrival of an *industrial growth society* is the perception of power in our relationship with nature, or having any connection at all.[21]

Take Shakespeare's understanding that nature is full of glory, mystery, and pleasure; a reading of *A Midsummer Night's Dream* with its magic and mischief plays with the innate power of forest life, the moon and the Earth rotating, and the human ability to capture some of it, for some of the time, at your own risk. The power of nature to bless or curse human activities looms inevitably in the background.

---

18. Carson, *Silent Spring*.

19. The Rowland-Molina hypothesis (1974) correctly correlated the use of CFCs to the degradation of ozone layer.

20. Meadows et al., *Limits to Growth*.

21. Macy, "Great Turning," 1:24—2:07.

This primal experience is still faced with tele-reality shows of survival in the wild, even with no clothes on, or of homesteaders "off the grid." The drama of individuals can be captivating as a graphic reminder of our roots and relationship with nature: that we are, and always will be, dependent. Primal experience also has a selling point; a new car is more attractive if it battles successfully through a storm of angry trees, snow monsters, or mud trolls. Beware, it's dangerous out there! Sadly, the extreme weather of climate change evokes this fear again, with many companies ready to sell you safety against the old enemy.

You can always place yourself at risk to natural forces, but more often human beings are in control and as never before; in an influential essay, Lynn White described the conflictual relationship between human civilization and nature.[22] Nature was to be held back and conquered. The colonial era of exploration and exploitation set European culture over against *everything*: foreign human cultures and nature alike.

The near extinction of bison, relatives of buffalo, in North America is a good example of these shifting power relations. Bison are emblematic of the Wild West. The language and portrayal of this history included much deceit. Just as the conflict was brutally violent and unjust for the Amerindian tribes, so too for bison, because white men knew they were the tribes' economic and spiritual foundation. The photos of astounding piles of bison bones framed the contrast between the integration of a hunted bison in the life of a tribe, when all of it would be useful, and the thrown-away carcasses of white man's violence: a double genocide, shocking and shameful.

Happily, the efforts to save the bison from extinction have been successful: a viable remnant population is present and growing in the USA and Canada. Restitution to tribes grows within limits too. While Green "founders" like Henry David Thoreau and John Muir might have found hope in this sadness, the catastrophic collapse of such a top species is an eloquent description of failures. There is a bucket full of tears for such atrocities towards one another, and nature, which continues with men, and violence, and perceptions of power. Tears have watered life.

Hope is a vital part of the Green political movement. Can you tell me of worthwhile decisions and actions arising from people who lost hope? Hopelessness is a feature common to many acts of extreme violence. So nonviolent direct action, a growing feature of Green politics, is far from hopeless. Be alert to misinformation: activists of Greenpeace and latterly Extinction Rebellion, and even local birders in Montreal, have been labeled ecoterrorists! Yet it was French secret service agents who exploded a charge

---

22. White, "Historical Roots."

to sink the Greenpeace ship, Rainbow Warrior, on July 10, 1985. This took the life of photographer Fernando Pereira. Nonviolence is a key principle in Green political parties.

The party politics of the Green movement has reached the consensus of a Global Greens Charter.[23] Six key principles place human needs alongside the environment: ecological wisdom, social justice, participatory democracy, nonviolence, sustainability, and respect for diversity. Concern for biodiversity includes a concern for human diversity; the principle of participatory democracy reflects the knowledge of how ecosystems flourish. Just as shifts in *human-nature* power relations are part of the Green movement, so *human-human* power relations are questioned.

I pick out the profoundly good news of some national governments to set human happiness as their new goal of success: the senselessness of gross national product as a standard of "greatness" has been recognized, notably by the New Zealand budget declaration, May 30, 2019.[24] This belongs with the salvation of Jesus Green (ch. 7).

In sum, the Green movement has grown in scale and form, to make it a multifaceted, multilevel, and multinational phenomenon. Each level of this movement is important and relates to the others: from neighbors who notice the value of a tree; to municipal bylaws on light pollution; to regional, provincial (state), and national policies; and international jurisdictions. It is a global movement within which each of us can participate. With it we can influence the future and breathe.

CHOICE IN WORK AND PLAY

Two famous sayings of Jesus resonate with the choice to be Green. Neither is very reassuring but reflect a teacher who confronted as well as affirmed.

> First take the log out of your own eye, and then you will see clearly to take the speck out of your neighbor's. (Matt 7:5)
> Be perfect, therefore, as your heavenly Father is perfect. (Matt 5:48)

These warn us we are bound to be hypocritical, but to be Green is a full-time calling, and just as John Wesley wrote, "Christianity is essentially a social religion," you cannot *be* Green alone.[25] I look for choices that cohere

---

23. The Global Greens Charter was adopted in Canberra in 2001, then updated Dakar in 2012 and in Liverpool in 2017.

24. Samuel, "Forget GDP."

25. Sermon 19, "Upon our Lord's Sermon on the Mount," discourse 4, in J. Wesley, *Forty-Four Sermons*, 237.

rather than contradict myself, but these are choices about what is beyond me: not just me, myself, and I.

We are especially reliant on good information, which requires a level of trust as well as questioning. The challenge is the volume of information available. Rather than offer advice on what to do, I highlight the active aspect of choice. The choice of the source of your information, your diet, a career (if you are lucky to have one), and the lifestyle with it takes energy. Choice circulates around Jesus' challenging sayings, despite our different era and context.

Perhaps the first word that comes to mind for Green, as care for the environment, is "recycling." To animate our church table at an Earth Hour event, we displayed the classic Green credo of "repair, renew, recycle" and invited people to add others, like "recognize, restore, rethink, refrain." The list reached twenty more, with great conversations.

Take this thinking to how we *work and play*. Both are fundamental to meaningful life. It may be your nature to take life as it comes. Do you work within a system that is uncool, anti-Green? Green is truly challenging. Perhaps your work is humdrum, just as school was a chore and life happened elsewhere. There is a level of dissatisfaction in any employment; to introduce Green to this risks upping the frustration. To be Green in *play* is equally problematic. Isn't leisure meant to be carefree pleasure? Is your passion for a sport like Formula 1 racing questionable? "My play isn't serious! I want to have the freedom to choose how serious I am in my leisure time." How do you spend, on what is nonessential? How can Green be part of that?

It is impossible to extricate ourselves from compromise, just as being rich in Jesus' days was suspect in his view. It comes down to systems, and we need each other to change them.

Two phenomena are good examples for the caution I raise over individual successes to be Green: the persistence of "fast fashion" despite its pollution and throw-away consequences, as well as documented horror stories of worker exploitation; and the popularity of oversized domestic vehicles. Both grow because of our addiction to pleasure. We are hardwired with this vulnerability, and businesses know it. GreenBiz founder Joel Makower blogged his twenty-first-century assessment:

> I'm not quite ready to proclaim green consumerism dead (though I can't honestly say it's ever been alive and well). There will always be a small corps of true-blue green consumers ready to vote with their dollars—at least for some products. But my 20-year-old premise—that a relative handful of committed

consumers will transform companies and markets—hasn't really panned out, though I still believe it to be true.

What will green consumerism look like over the next decade? Will we be celebrating or mourning green consumerism when Earth Day 2020 rolls around? And if the former, how will we have gotten there? I welcome your thoughts.[26]

Earth Day 2020 has come and gone, and Green consumerism is a bag of success and failure. Voting with your dollar has the same pitfalls as regular voting, to choose from options that may not represent you or be what they're claiming to be at all.

Milton Friedman, a leading economist of the sixties, was famous for his phrase "the tyranny of the majority," as he wrote about the limits of democracy (so that there are always losers) and tyranny being less likely in a free market, as it encourages diversity. Not so, says Joel Waldfogel. The way markets function means I can easily find myself left out: for a long time, this meant that vegetarian meals and organic veggies were hard to find: "You will find products that suit you only if enough others want the product."[27] The positive side of this is for neighborhood initiatives. To shop local influences the market: local pockets of Green people can improve the choices available to them and their neighbors.

I am skeptical. A cartoon shows a mother and daughter buying sunglasses: while the daughter takes a selfie to see how she looks, her mother frustratingly screams, "Will you just look in the mirror?" The small screen is "bigger" than life. We are in the first and second generation of social media impacts, through smartphones, and how this affects behaviors and self-identities. When can there be training and licenses for use of cell phones in public?

I will tell you what I bought and what I boycotted: movements can form faster than ever, influencing the choice of millions.

Can you admit your freedom to choose is an illusion? Monty Python were spot on again. In the film *Life of Brian*, as Brian is faced with a sea of admirers who think he is the Messiah, his every word is counted. He tries to persuade a crowd it's all a big mistake, he is not Christ. "Listen!" he pleads, "You are all individuals!" And, as one, the cry comes back: "We are all individuals!"[28]

Individual twenty-first-century adolescents suffer far more identity anxieties than I did in the 1970s. New freedoms of expressions online still

---

26. Makower, "Green Consumer, 1990–2010," paras. 15–16.
27. Waldfogel, *Tyranny of the Market*, 3–4.
28. Jones, *Life of Brian*.

come with the need to be validated, especially by our peers. I want to be me, but I still need to feel accepted. The result is a dynamic and rapid sea of "choices" that mimic herding. I can hear music from Handel's *Messiah*, "We like sheep have gone astray." What do you think lies behind the vocabulary of "influencer" and their "followers"?

The limits to individuality are also socioeconomic. Will I be as successful as my parents? How can I escape poverty? I do not want to suffer the way my mother or grandmother did. Our career choices, if we are lucky to have them, are in the shadow of our upbringing.

What were your choices, family circumstances, and how do you play through these consequences today? Our starting points limit our freedom as a social DNA.

If you are against the death penalty, I presume you would not accept being an executioner; a vegetarian would not work in an abattoir, a pacifist never in the military. Are there some jobs that are un-Green? Is Green a declaration to refuse employment with certain companies or even professions entirely?

Most of us are in much trickier situations. Life throws compromise all the time. What does the ecologist do if offered a contract for a biodiversity assessment of land suited to development? Many environmentalists serve government departments whose policies can be contradictory.

Do I apply to work for a multinational to pay off my student loan? Do the compromises I make with my first job continue my life long? I might end up driving the bulldozer that clears virgin forest or takes tar sands tailings. Does the appeal of the plum job I want include regular conferences round the world, with huge environmental impact? When do I make a choice of employment that is not just about the paycheck? How unfair and bourgeois to consider choice of employment when so many of us are simply glad to have work and income. We may have "failed" in school and taken on the label of stupid or inadequate; our childhood wounds predispose us to a life of dependency. We are excluded, so ready and glad to join any system, for daily food and stability in which to raise a family. This dependency is seen in some of the most dramatic ecological crises, where small farmers destroy pristine forest, or poachers take out endangered species, or First Nations are divided over an offer of "land for pipelines." It explains how Brazil was so deeply divided in the 2022 presidential elections.

Choice is wedded to compromise, but Green choices intentionally change where we put our time and talents, spirit and energy; how we spend our money. I seek information from organizations I can trust, because they understand the living world: they are engaged in social change and may be large enough to resist commercial or political pressure.

This information tells me one of the best and simple decisions we can make is *to eat less meat*. Green is to find special pleasure in eating your greens *before* you buy an electric car or think "not flying" is *the* solution! "Business as usual" predicts a growth in meat demand that makes the two-degree climate target impossible. It is as bad as that.[29] A focus on the threat of meat consumption reveals how individual choice and corporate action are two sides of the same coin. Regulation as well as reduction is key. The political will to pass regulation relies upon enough people whose actions say, "Make it happen!"

The humdrum ensemble of daily life, families, schools, power supplies, contracts, reliable exchanges of goods and services, landmark choices, and small daily ones make up our work and play.

As this book unfolds, I hope you will discover Green, as choice, is lightened by Spirit, with new pleasures and goals. This will not be fulfilled through individual choice alone: but not to live differently, and argue others must, has no sense or authority. We are bound to choose Green. Let the energy and the movement of Green empower your choices; participate and flow. Tell your friends these trends are Green, it's natural, like looking in the mirror. We live to let others smile at what they see.

CONSCIOUSNESS: DREAMS AND NIGHTMARES

One of my favorite musicals is Rodgers and Hammerstein's *South Pacific*, especially the 1958 movie by director Joshua Logan. It follows the fortunes of American soldiers and Pacific civilians in the Second World War. Two romances across age and culture raise the classic question of romantic love that challenges social expectations. A young US lieutenant Joe Cable sings, "You've got to be carefully taught!"

> You've got to be taught before it's too late.
> Before you are six or seven or eight.
> To hate all the people your relatives hate—
> You've got to be carefully taught!

Joe Cable has fallen for a Polynesian girl and knows his true love would not be accepted back home; worse, he knows he lacks the inner strength to confront it. His dream of love has become a nightmare; but at least he knows he was not born that way.

This is our Green experience too. The biophilia that can entrance us brings hurt and sadness. Our upbringing influences our choices for work

---

29. Böll, *Meat Atlas*.

and play more than we realize. We have been carefully taught what is progress and what things are worth.

Summer 2019 included a shocking heat wave across Europe and the meltdown of Greenland glaciers. Imagine what impression this leaves upon those aged six or seven or eight. Children absorb the concerns of their parents, the emotions of news headlines. We have to acknowledge a sense of doom in the young that is new and important. Is there understanding to go alongside the rising alarm? Climate change denial is being carefully taught at home, but what about at school?

Suppose something accurate is being taught: that the natural earth as we know it, is being destroyed by human activities. This a journey from a dreamy wonder for the living world to a nightmare of its destruction and the birth of a guilty conscience. If I were to have a conversation with a bison, eye to eye, I would be ashamed.

Our guilt mounts year to year, and often our reactions are to sweep it away and justify ourselves, but deep down it rests. Guilt is a toxic part of the environmental nightmare because it undermines conversations and change. Do I want my shame to be in the light?

Let me take you to a Christian hymn for this guilt, "Ah, Holy Jesus." It is a Good Friday classic, to the tune "Herzliebster Jesu" by Johann Crüger (1598–1662).

> Who was the guilty? Who brought this upon thee?
> Alas, my treason, Jesus, hath undone thee!
> 'Twas I, Lord Jesus, I it was denied thee;
> I crucified thee.[30]

Of course, this is impossible unless our imagination gives insights to reality. It is true only to the extent that the drama of the Jesus' story is universal. I offer it because guilt and Green go together, and some antidote to guilt is needed, alongside the consciousness that is Green.

We do not crucify people these days; we shoot one another. Guns and Green weave together powerfully. They are part of the life of one of America's most famous painters, John James Audubon (1785–1851). Audubon's book *Birds of America* has been sold for record sums at auction because it combined science and art in ways that inspired new appreciation and knowledge. Audubon wanted to be precise and created a system of drawing birds to exact scale. The largest available paper size, double-elephant, was approximately forty inches by twenty-nine inches, so he had to paint the largest subjects in strange postures. Audubon's accuracy was not just in size,

---

30. Heermann, "Ah, Holy Jesus," stanza 2.

he painted the truth of the birds by setting them in nature: eating the berries of their choice, cavorting, seizing prey, male and female, youngsters and chicks. You know he understood his subjects, or even misunderstood them when making false assumptions.

Then, I discovered Audubon killed as many birds as anyone of his generation. He was a hunter as well as ornithologist and had no consciousness that his activities contradicted themselves. Out of the 425 species he painted, five have since become extinct, including the great auk, the passenger pigeon, and the Carolina parakeet. A sixth, the ivory-billed woodpecker, teeters on the brink. The collapse of the passenger pigeon population is even more incredible than the near extinction of the bison. The descriptions of their movements are extraordinary. Audubon estimated one of the mega-groupings he witnessed, flying uninterrupted for three days, to be twenty-five billion birds.[31] The impact of their arrival was devastating on crops and the possibility their numbers would collapse unthinkable. It was another era, and only recently has the biology of species of mass movements been understood. Massive flocks have their specific adaptations, so that even when shooting them was recognized to be a problem, the huge but smaller groups that remained were unsustainable. Audubon was alive in the first half of the nineteenth century, but it took less than a hundred years for the passenger pigeon to be shot to extinction.[32] Protective legislation was introduced ineffectively or resisted entirely during the second half of the 1800s. Massacres of over fifty thousand birds per day continued for nearly five months of 1878 in Pennsylvania. Slaughter continued despite the evidence of declining populations. Collectively, there was no will or ability to change behavior in the use of guns.

As I write, there have been two mass shootings within twenty-four hours in the USA (August 2019), and the nation appears as divided as ever over easy access to deadly weapons as a key factor. If human-human violence cannot be adequately resolved in civic life, then what chance do we have to reduce the violence done to other species? We can become very pessimistic as to our chances, or even our merits as a species who have so much power within our control.

In fact, this trip through the use of guns, species extinction, and human violence shows we do not have control over our collective behavior, and the impact of culture in determining behavior is still underestimated. We ignore being carefully taught.

---

31. Souder, *Under a Wild Sky*, 154.
32. Fuller, *Passenger Pigeon*, 50–69.

How else can we explain behavior in Quebec, where environmental impacts are daily headlines and hundreds of thousands make public demonstrations, yet the sales of energy-hungry jeeps and oversized vehicles are rising. Arguably the majority are still dreaming of themselves in fantasies of a limitless road movie.

All this feeds a reasonable basis for ecological despair: eco-grief, a Green consciousness that will deepen sharply. One crisis will lead to another, with the inertia of consumer-driven economies and the lack of political courage to take up models that have been around for three generations.

But what is the point of my speculation upon catastrophe? My opinions are not fully informed. Rather, consider how remarkable social change has succeeded and the odds of this success, such as abolition of the slave trade and slavery. The film *Amazing Grace* tells some of the story.[33] In the UK it was a drawn-out battle, and won, at first, only by a sort of anti-French trickery to ban the slave trade under other flags (Foreign Slave Trade Abolition Bill, 1806). Abolitionists knew the majority of the British slave trade was under foreign flags. The British Parliament voted to abolish the slave trade a year later in 1807.[34] For twenty years the arguments for British jobs and businesses held sway against arguments for abolition, despite the basic moral offensiveness of it all. This at a time when Christianity was the national religion. There are differing opinions on how this change came about. Less known is the impact of resistance and rebellion by slaves themselves, noting the Haitian uprising of 1805. The complexity of this parallels our Green concerns; economic, political, and moral actions were all at play. It will be this: to struggle against a seemingly intractable resistance to change, conscious and often unconscious.

Beware of false prophets: Stephen Pinker argues that Enlightenment consciousness—reason, science, and humanism—is working to bring the sort of hopeful progress we need.[35] I hope you are being enlightened in reading this book, but he means the interdisciplinary perspective that was classically found in the Renaissance. There is an important debate on the impact of some Enlightenment thinking: why would Newton be preferred over Einstein? *Cogito, ergo sum*, Descartes' famous phrase, "I think, therefore I am," is inadequate and outdated. It is dangerous to encourage human self-understanding without any reference to the natural world. Something more than Enlightenment consciousness has emerged: systematic consciousness; the collective consciousness that the Internet has made possible,

---

33. Apted, *Amazing Grace*.
34. UK Parliament, "Parliament Abolishes Slave Trade."
35. Pinker, *Enlightenment Now*.

with the media coverage, blogging, tweeting and clustering of interests that is global, rapid, and interactive. If this is a second Enlightenment era, it includes this additional global consciousness that means hopeful change is possible, faster than ever; with a "we" as well as an "I"; it includes the selfie of us sharing planet Earth and it is a basis for "yes, we can."

In religious terms, the antidote to guilt can be atonement: practical actions that pay back the damage of human activities. Guns again: Aldo Leopold (1887–1948) graduated from one of the first forestry schools, at Yale University, and joined the US Forestry Service (1909). He was paid to cull wolves as part of park management program. Then one shot of a wolf with cubs changed him:

> We reached the old wolf in time to watch a fierce green fire dying in her eyes. I realized then, and have known ever since, that there was something new to me in those eyes—something known to her and the mountain. I thought that because fewer wolves meant more deer, that no wolves would mean a hunter's paradise. But after seeing the green fire die, I sensed that neither wolf nor mountain agreed with such a view. Since then, I have lived to see state after state extirpate its wolves . . . I have seen every edible bush and seedling browsed . . . and the starved bones of the hoped for deer herd. I now suspect that just as a deer lives in mortal fear of its wolves, so does the mountain live in mortal fear of its deer.[36]

Leopold had grown up in Iowa near the Mississippi and developed an affinity with nature that let him see it as a community. He had the courage to join those opposed to culls of wolves, and where the practice changed, wholesale effects were seen on ecosystems. This demonstrated one of the first recognized examples of trophic flow. The effects of the activities of a top predator are key to the overall balance of an ecosystem. In the case of the wolf, the predation upon herbivores, deer and moose in particular, means they have to keep on the move and grazing in one place is limited. More saplings of birch and ash reach maturity, and the renewal cycle of the forest can continue. Removal of wolves therefore led to overgrazing and thinning of the forest, loss of biodiversity and soil retention. It undermined a renewal of the entire forest ecosystem. There was less life instead of more.[37]

---

36. Leopold, *Sand County Almanac*, 129.

37. Direct damage is done routinely by humans. Paul Agnew, a forester in British Colombia, described to me the impact of clear-cut hillsides; the death of ground plants and washed away soil, to expose "rocks like skeletons and scars on the body of the earth" (conversation with author, January 2024).

If human beings, as the ultimate top predator, believe that our actions do not have similar, wholesale impacts, we deceive ourselves. Twenty-first-century Enlightenment is a respect for pristine nature and the dynamic relationship we have with living things, at all levels of human societies and cultures.

We are responsible, like it or not.

Green consciousness is about finding a new awareness that is neither a fantasy of dreams nor a nightmare of despair. It is a consciousness of praxis, actions based on valid understandings, which are hopeful because they are part of a shared, unprecedented shift in human behaviors. Green consciousness includes the knowledge, so far as we have it, of the processes of life. It bases human choice on the foundations of the realities of living systems, as a positive response to "overshoot."

Although I have offered a mere sketch of what is Green, the combination of energy, movement, choice, and consciousness is more than the sum of the parts. It lets us reconsider what we were carefully taught and change our aims.

Chapter 3

The Green Jesus?

JESUS KNEW NATURE

What is the difference between the Green Jesus and Jesus Green? This chapter is for those who would not notice and those who will want me to get to the point. The answer comes only if we work through the surface of things. This is how I treat my massage clients, to work deep spinal muscles by effleurage *before* working on a tight spot. We work through the body of Scripture from a surface reading to deep structure. A body analogy also encourages more than analytical thinking: the combination of perspectives in this Jesus mosaic will count.

You are right to expect to form your own opinions about Jesus. The four Gospels of the New Testament are not spiritual equivalents of computer programs, to deliver certainty about God: they are provocative storytelling with a purpose. They tell us Jesus provoked a range of reactions from ordinary people, and so it continues!

Proof of the Green Jesus is as rare as water turned to good wine at a wedding (John 2:1–12). But Jesus was close to nature, and it shapes his teachings. To enter his story is to act as we see in the film *Avatar* with veteran Jake Skully.[1] We can "meet" Jesus in at least two worlds: the ancient and the contemporary.

Some Bible verses suggest *Jesus was aware of nature*, of itself and in human affairs. As is typical in a study of Scriptures, once we have a different angle, we see what has not been seen before. Where there are similar passages in other Gospels, I list these as parallels (//).

---

1. Cameron, *Avatar*.

Matthew
And when Jesus had been baptized, just as he came up from the water, suddenly the heavens were opened to him, and he saw the Spirit of God descending like a dove and alighting on him. (3:16//Mark 1:10; Luke 3:22)
Six days later, Jesus took with him Peter and James and his brother John and led them up a high mountain, by themselves. (17:1)

Mark
He was in the wilderness forty days, tempted by Satan; and he was with the wild beasts; and the angels waited on him. (1:13)
He woke up, and rebuked the wind, and said to the sea, "Peace, be still." (4:39//Luke 8:24)
Jesus said to him, "Truly I tell you, this day, this very night, before the cock crows twice, you will deny me three times." (14:30//Luke 22:34)

Luke
When he had finished speaking, he said to Simon, "Put out into the deep water and let down your nets for a catch." (5:4)
"Go into the village ahead of you, and as you enter it you will find tied there a colt that has never been ridden." (19:30//Matt 21:2)
He came out and went, as was his custom, to the Mount of Olives. (22:39//Mark 14:26; Matt 26:30)

Other verses suggest that *Jesus made use of nature to make his points*:

Matthew
"For he makes his sun rise on the evil and on the good and sends rains on the righteous and on the unrighteous." (5:45)
"Look at the birds of the air." (6:26)
"Consider the lilies of the field." (6:28)
"You will know them by their fruits. Are grapes gathered from thorns?" (7:16)
"See, I am sending you out like sheep into the midst of wolves; so be wise as serpents and innocent as doves." (10:16)
"Are not two sparrows sold for a penny? Yet not one of them will fall to the ground apart from your Father." (10:29//Luke 12:6)
"Listen! A sower went out to sow." (13:3-9, 18-23//Mark 4:3-9; Luke 8:4-18; 13:18-21)
"The kingdom of heaven may be compared to someone who sowed good seed in his field; but while everybody was asleep, an

enemy came and sowed weeds among the wheat and then went away." (13:24–30)

"The kingdom of heaven is like a net that was thrown into the sea and caught fish of every kind." (13:47)

"If you had faith the size of a mustard seed, you will say to this mountain, 'Move from here to there' and it will move." (17:20)

"Again, I tell you, it is easier for a camel to go through the eye of a needle than for someone who is rich to enter the kingdom of God." (19:24//Mark 10:25; had Jesus experience of leading camels? He makes us smile.)

"All the nations will be gathered before him and he will separate people one from another as a shepherd separates the sheep from the goats." (25:32; goats and sheep resemble each other in Palestine.)

Luke

He also said to the crowds, "When you see a cloud rising in the west, you immediately say, it is going to rain and so it happens.... You hypocrites! You know how to interpret the appearance of earth and sky, but why do you not know how to interpret the present time?" (12:54–56)

Then he told this parable. "A man had a fig tree planted in his vineyard." (13:6–9)

He said therefore, "What is the kingdom of God like? And to what should I compare it? It is like a mustard seed that someone took and sowed in the garden; it grew and became a tree and the birds of the air made nests in its branches." (13:18//Matt 13:31–32)

"Jerusalem, Jerusalem, the city that kills the prophets and stones those who are sent to it! How often have I desired to gather your children together, as a hen gathers her brood under her wings, and you were not willing!" (13:34//Matt 23:37)

So he told them this parable: "Which one of you, having a hundred sheep and losing one of them." (15:3–7)

The Lord replied, "If you had faith the size of a mustard seed, you could say to this mulberry tree, 'Be uprooted and planted in the sea,' and it would obey you." (17:6)

"There will be signs in the sun, the moon and the stars, and on the earth distress among nations confused by the roaring of the sea and the waves." (21:25)

"Look at the fig tree and all the trees; as soon as they sprout leaves you can see for yourselves and know that summer is already near." (21:29//Matt 24:32)

There are verses in John's Gospel that draw on nature, but they are overlaid with a symbolism that puts us at a further distance from Jesus himself: the bread of life (John 6:35), the light of the world (John 8:12), the good shepherd (John 10:14), and the true vine (John 15:1).

You will not find any verses that instruct us to be Green. No one would have believed it possible to destroy nature as human activity does today.

We can even find some contradictory evidence: a fig tree is cursed (Mark 11:13–14, 21; Matt 21:19–21) and pigs drown in the sea (Mark 5:1–13). The descriptions of a final judgment also leave the hearer in doubt of the safety of creation (Mark 13:8–21; Matt 24:4–36; Luke 21:8–36). But later apocalyptic descriptions balance that with the promise of a new heaven and a new earth (Rev 21 and 22).

No one has dared to suggest Jesus was Green, because this is to lift Jesus into our contemporary concerns and put him back in Scripture, denying the process. However, the more I have considered what is Green with the studies of the historical Jesus and the New Testament, the more I find Jesus would be Green today. Jesus calls me to be Green, through patterns of thought and concern: he may be faint Green, but this is enough. Nature was more present to him and his society than to ours, in the industrialized West. Our nature loss is not to be projected on to Jesus of Nazareth.

Jesus took nature as part of his reference to God, but he had other priorities: healing and teaching that the realm of God was close. He taught about forgiveness of sins when forgiveness through the temple of Jerusalem was an experience of exploitation. Many believed God would bring an apocalypse so the urgency of offering wider forgiveness made sense. Jesus was conscious of the material, social, and spiritual crisis of the crowds following him. He was concerned for the worst victims of the crisis: cripples, the spirit possessed, lepers, and the sexually exploited. He charged religious leaders with responsibilities they had abandoned and closed to others (Luke 11:52; Matt 23:29). Jesus drew on nature as a resource for the human drama that dominated his mission.

However, an apparently weak Jesus is a strength. Faint Green can become deep Green if our understandings of Jesus and Green grow and change.

A QUEST FOR THE GREEN JESUS

Like the search for the lost ark of the covenant, a quest for the Green Jesus will not go away, but it is bound to be disappointing unless something more experiential can take place in me and you: a living tension, through the novelty of the past and the demands of the present.

One analogy would be how to become a coffee connoisseur. A TV reality show, *Dangerous Grounds*, followed Todd Carmichael, owner of La Colombe Torrefaction, and his passion to deliver rare coffees to wealthy clients. Carmichael did this through a love for people, culture, and common friendship. I thought I took coffee seriously already, but there are so many varieties of beans, and I had not realized the "life-or-death" dependency of communities where they are grown. Each community has its history, which makes up the story of global coffee production. Now I take more interest in the variety of fair-trade coffees I can buy, and I am less likely to let someone else tell me what makes a good coffee. There is a past and many locations that inform my appreciation of each present moment as I sip it.

Like coffee hunters, Jesus seekers must appreciate the ground of things. This is more than wordplay. The raw form of ancient writings are strange to us: consider this, from a parchment dated 280 CE:

> Πάντα δι' αὐτοῦ ἐγένετο, καὶ χωρὶς αὐτοῦ ἐγένετο οὐδὲ ἓν ὃ γέγονεν.
> Panta di' autou egeneto, kai chōris autou egeneto oude hen ho gegonen. (John 1:3)

The second line is a transcription of the first, from Greek letters into Arabic English. Even if you know the verse by heart—"All things came into being through him, and without him not one thing came into being"—it is difficult and strange to read this transcription out loud. I chose a verse that speaks to Green concerns, but the whole New Testament demands this exercise of translation, which is normally hidden. Uncertainty and strangeness are lost in the production of Bibles for mother tongues around the world. I take us one layer beneath the surface.

Scholars estimate John's Gospel was written around 120 CE. The process of production to bring it to English-speaking Christians has passed it through more hands than coffee beans. Since 1611 the King James Bible made Christian Scriptures accessible to English readers. It was taken as the literal word of God; many still associate the strength of the Bible with old-fashioned English. This is to mistake the commanding artistry of the language as the text. The King James is a product of several translations, from Greek into Latin and then into English, and the Greek relied on texts that were later than those available to scholars today. Errors and faithful interference take place at each step of production, distancing us from the original.

This weakens an easy reading of the Bible, but we can lay some foundations to restore some of its authority.

First, realize there is an ongoing global dialogue between congregations and universities. Communities of faith read the Bible daily and weekly as fuel for faith whereas scholars know how present-day convictions and cultures may seriously distort its reading. A dialogue is necessary, but this began only in the 1800s when studying the Bible as just another book caused shock waves. Today, the academics of biblical studies may be secular, radical, or conservative, all engaged in free research and with their personal convictions. Yet the faith world of Christians did not collapse when deeper questions gave surprising answers.

Second, thinking historically is one of the foundations that informs biblical study. It is not obvious for some, just as many find ecological thinking troublesome. Take the famous Henry Ford: "What do we care what they did 500 or 1000 years ago? . . . It means nothing to me. History is more or less bunk. It's tradition . . . the only history that is worth a tinker's dam is the history we make today."[2]

This is passable as an opinion if you make motorcars, but startling if we are dealing with the story of Jesus. Nevertheless, a renowned theologian, Karl Barth (1886–1968), downgraded the importance of history. For Barth, history was suspect. The atrocities of warfare between Christian nations demanded a grim reassessment. He insisted upon the radical interference of God in Jesus Christ for human salvation through faith alone. Barth wrote inspiringly, yet his perspective was itself due to the history of genocide and blitz bombing. Contemporary Christians who deny history affects reading the Bible can be said to be reactionary and inconsistent. They do a disservice to Jesus. Perhaps they can simply be asked, "Do you read Greek?"

Historical thinking is a bit like breathing: it will never cease, but I had to be woken up to realize it. The classic book parodying British—sorry, English history—*1066 and All That*, plays with the notions of history as mere recollections of facts.[3] History "starts" with the last successful invasion of Britain, by William the Conqueror in 1066. The history I learned in my schooling was sadly like this, but in training for ministry I began to realize the dynamic nature of the subject. New Testament history scholar Patrick Henry contrasts Martin Luther (famous for his declaration of "Here I stand, I can do no other" after nailing his statement of faith to the cathedral door) with a character from the musical *Fiddler on the Roof*, Tevye, who ponders

---

2. Butterfield, "Henry Ford," para. 6.
3. Sellar and Yeatman, *1066 and All That*.

life's choices, on the one hand "this" and on the other hand "that."[4] Tevye stands for most of us. Luther's stance was exceptional. But Henry claims both sorts of human reactions are helpful in understanding the New Testament: this is an important realization, that "historical thinking is not an exclusively intellectual activity."[5] Even the recollection of "facts" becomes an exercise that reflects myself. What do *I* remember? He concludes "that the work of the historian is an extension of the human faculty of *remembering*." I have a poor memory, but what I recall, and how I tell it, is also very personal.[6]

There are clues to Jesus being a historian, when he said goodbye to his friends with a ritual of bread and wine as his body and his blood: "Do this to remember me" (1 Cor 11:24, 25). This tangible basis to being Christian is key to how it claimed my youthful devotion. The physicality of Jesus let my youthful passion connect with his passion, and I discovered faith is as dynamic as history. I had no idea that it matched up with theologian Paul Tillich's major work, *Dynamics of Faith* (1957).

I have accepted therefore that history can grasp the past only in relative terms, as a continual process of reflection and change. Rather than weakening history, this leads me to question and care about my experience of reality in general. Did you enjoy the film *Groundhog Day*? The story takes us in a repeated loop of twenty-four hours, and plays out a game of "consequences" with the question of "what if" versus a "so what."[7]

After a surface reading, we must move away from the expectation that the Bible must make sense to us if we simply open it and read it. Coffee is complex! New Testament studies are richer and continue to develop. The quest for a Green Jesus takes us into unfamiliar lands.

KEYS TO THE PUZZLE OF JESUS

Have you ever been involved in a drama that became a news story? What you know firsthand and what is reported in the news can be so different. Even when it is a fair account, it shows how journalism must leave things out.

Have you been to court? The perspectives of accused and accuser are explored with witnesses; each may in honesty tell the truth, yet that truth is limited according to each. There is a summary by the judge, trained to

4. Stein, *Fiddler on the Roof*, 73, 124.
5. Henry, *New Directions*, 29.
6. Henry, *New Directions*, 35; emphasis original.
7. Ramis, *Groundhog Day*.

clarify this complexity to encourage a correct decision. Complexity lies with the original. Simplifying things can be cheating ourselves of the truth.

So, studying Jesus becomes an enigma that scholars such as Henry sought to unpuzzle by three contrasting couplets:[8]

Unity and diversity: You may perceive the discrepancies among the four Gospels as a weakness, in fact they were part of the approval of Christian diversity by the early church. This was the judgment of Bishop Irenaeus (author of the Irenaeus Creed) in the second century CE. He argued there were four Gospels to correspond with the four regions of the known world. The authorization of the four Gospels was certainly political, but the diversity it guaranteed was then used to unify Christianity against heretical criticisms. Given the diversity that we find in the church today, it is important to realize the early Christian leaders also had the question of unity and diversity in their minds. They asked, what holds the church together? It is still a very contemporary question in a world of glaring contrasts.

Continuity and discontinuity: Jesus' Holy Scriptures were the words of the Old Testament. The fact that the *New* Testament was written points to a break from this profound tradition. But as a Testament, it claims continuity of some kind too. The tension between continuity and discontinuity has been present from early days. An argument for discontinuity came from Marcion, a brilliant gnostic Christian leader in the second century CE. He demanded all the Old Testament references in the New be rejected. A contemporary example is by Jacob Neusner.[9] Neusner claims little interaction between Jews and Christians took place and that dialog requires recognition of the separate integrity of Jewish and Christian religions.

We can find an echo of this tendency with the view that God's nature is different in the two Testaments. God is a God of wrath in the Old and a God of love in the New; but attention to both Testaments quickly shows they are multilayered complex collections of books, and the portrayal of God is similarly rich and irreducible in each.

Then our own contexts affect how the Bible is read, with the God of love being favored in economically secure cultures and the God of justice and liberation prominent among the oppressed. Rather than being a radical break with Jewish identity, Christian faith is better portrayed as an unprecedented development.

The case for continuity has rested on the first Christians being centered in Jerusalem, the Old Testament predictions of a Messiah (Christ), and the use of allegory to find Christian meaning in the Old Testament. But the

---

8. Henry, *New Directions*, 41–69.
9. Neusner, *Jews and Christians*.

best explanation comes from the complexity of Jewish life in the times of Jesus: a small sect would be tolerated at its very beginning simply because it went unnoticed in the prevalent tapestry of Jewish faith practice. Four independent sources describe this Jewish diversity: the New Testament, Flavius Josephus (first century CE), rabbinic texts of the second century CE, and the Dead Sea Scrolls.

Biblical language and meaning: We have seen how strange ancient Greek parchment looks to most of us. This is vital to realize, as the study of Jesus is shaped and limited by this written medium and when and why it was created. Once translated, a literal meaning can be ambiguous, and this begs us to place a passage or verse in some context. The church disseminated the New Testament across centuries and the world. So, we consider the original communities for whom the documents were written and the authors who intended to share their understandings and knowledge; this knowledge was drawn from firsthand experience, living witnesses, their cousins, the stories handed down in families, by speech not written word.

This is a sense of history as faith history. We have texts with a purpose, resting upon events. The text makes meaning out of those events, a meaning the author judged met the needs of the church. When the church preserved the four Gospels, it was an admission of the importance of diversity. There was a respect for faith history that gave conflicting factual details, the order in which things happened or fashion in which they were said. Some of this respect was an admission of the importance of the integrity of the written words.

Hold these three couplets together: unity and diversity, continuity and discontinuity, text and meaning. They have, in effect, exploded a naïve reading of the New Testament. You cannot go back. The role of the early church to create and then preserve the New Testament adds complexity and uncertainty, but somewhat disentangled, we have a reasonably clearer picture of Jesus than before. Albeit a very partial picture that we see from a limited angle.

I can ask you a pointed question to challenge you to wrestle with these claims. You have a choice between a vision through lenses that make everything blurred to the point of being uncertain about seeing objects and people, or a narrower but clearer vision that requires an effort to turn your attention to what you want to see. Which would you trust your life to?

Importantly, for this gospel of Jesus Green, being conscious of how the New Testament was put together allows us to appreciate that creative faithful retelling of the Jesus story was allowed.

Just as a painter must give a raw canvas the proper preparation and base colors, we must grapple with the historical Jesus before he can possibly be Jesus Green.

WILDERNESS FORMATION AND PROVIDENCE

Jesus did not have the choices we have today. We can live relatively cut off from nature. So much so, there is a new Green spiritual program called Wild Church that intentionally takes participants into nature.[10] We offered this at my church with a walk from residential comfort around the church to "immersion" in a small nature park, at the highest point of Westmount. To do this month by month in Canada is to experience the drama of the seasons. It also shows me how urban life distances us from the power and peace of wilderness.

Jesus' society was dependent upon rains and good harvests, weeds not choking wheat, seeds falling on good soil. The fact that Jesus refers to nature according to the Gospels is not sufficient. What else could he draw upon to make illustrations? What matters is the *use* he makes of nature in his illustrations and allegories, and of course we know this is almost always the use of the editor, one step away at least from the source.

But these points may be unfair. A love of nature is found in every generation: Why not in Jesus, thanks to Matthew, Mark, Luke, and John? If we cannot get further back than the Gospel writers, let us at least go to these limits and see what is uncovered.

The first three Gospels each tell us Jesus' faith formation was in the wilderness. What this means is different, however, than a "John Muir sense of wilderness," with nature's glory completing human needs for communion. Biblical wilderness is a setting of extremes, where faith counts for everything: the wilderness broke and made the Hebrew people. Celtic spirituality names it as a *thin place*, and anthropology as *liminal*. You do not enter wilderness and remain the same. For the Hebrew people, they received the law and covenant of God as indelibly as a branding. This is the common ground for the Synoptics' understanding of wilderness, and each Gospel makes further points.

Mark begins his story with an adult Jesus in the crowds of people drawn by John the Baptist. It is the first of the authorized Gospels. His style uses short phrases, for a short, action-filled Gospel with codes and urgency. There is no independent account of Jesus' baptism, but historian Josephus records John and his activities in his *Jewish Antiquities*:

10. Wild Church Network in USA and Canada; Forest Church in UK.

> Now some of the Jews thought that the destruction of Herod's army came from God, and that very justly, as a punishment of what he did against John, that was called the Baptist: for Herod slew him, who was a good man, and commanded the Jews to exercise virtue, both as to righteousness towards one another, and piety towards God, and so to come to baptism; for that the washing [with water] would be acceptable to him, if they made use of it, not in order to the putting away [or the remission] of some sins [only], but for the purification of the body; supposing still that the soul was thoroughly purified beforehand by righteousness.[11]

Mark quotes the prophet Isaiah to set John and Jesus in the wilderness: "See, I am sending my messenger ahead of you, who will prepare your way, the voice of one crying out in the wilderness: 'Prepare the way of the Lord; make his paths straight'" (Isa 40:3; Mark 1:2). John baptizes Jesus in the river Jordan and Jesus goes on to spend forty days there (Mark 1:9–15).

This tells us Jesus identified with the prophets, the law, and the renewal of his people. Natural elements are at the fore: water, heavens, and a dovelike Spirit who drove Jesus to the wilderness, with wild beasts and angels, to be tempted by Satan. It refers to Genesis, the origins, where Adam and Eve were also with the wild beasts; to Daniel in the lion's den; and to Elijah, who survived thanks to angels and birds (1 Kgs 17:6; 19:4–8).

Given the richness of these references, Mark tells his story like a Modigliani sketch uses a minimum of lines. Jesus had credibility. He fitted the prophetic predictions of his people, with the wilderness as the origin of their identity. Jesus' identity was born from extreme "out-of-civilization" immersion: nature was the cauldron that determined Jesus' confrontational mission of good news.

Mark's interest in the wilderness was political and ideological. Jesus worked from the periphery to the center. He demanded a relocation of the activity of God, to ordinary people. The realm of God was to be received like a child (Mark 10:13–16). He disturbed the priests and scribes of Jerusalem when he told a parable favoring a Samaritan, and challenged the notions of purity by eating and drinking with dishonorable folk. The wilderness was the ultimate periphery. Jesus retired to lonely places to recharge and reaffirm his spiritual and political message. Wilderness had authenticity, resonance, and power.

Mark is not wrong. To move anywhere between villages and towns was to cross wilderness areas, and he knows his audience remember wilderness

---

11. *Antiquities*, 18.5.2., in Josephus, *Complete Works*, 581.

has a huge role in Jewish faith identity: Abram was ready to settle down, but he responded and set out "to the land that I will show you" to become Abraham, the father of the nation (Gen 12:1). Joseph is cast by his brothers into a pit "here in the wilderness": the stage for Jewish Egyptian origins (Gen 37:22). Moses kills an Egyptian and runs for his life to a mountain, meets a burning bush and God, who sends him back to Pharaoh and conflict (Exod 3:2–10). The ultimate wilderness experience for Moses and the Hebrew people meant forty years of wilderness wanderings with the giving of the law. In the promised land, settling down brought cities, kings, and loss of faith and justice. Prophets rose up to call all to account, and wilderness served as their place of refuge, revelation, and restoration.

All these references are at play as Mark tells us Jesus was in the wilderness. How much of the forty-day wilderness experience and temptation of Jesus actually happened? Jesus knew these prophetic and Genesis stories. It was a reasonable faith practice for him to do something like what Mark describes. Let us settle on Mark's Jesus being made through nature and the Spirit of God.

In the Gospel of Matthew, Mark's description of Jesus and wilderness has details that portray Jesus as the new Moses. Jesus was in the wilderness *forty days and forty nights* (shorter than forty years, but still). The night evokes the reality of hardships, heat, and cold, then Matthew uses the details of Moses receiving the law (Deut 9:9) and Jesus fasting throughout; at the very end, three related temptations are centered on Jesus' well-being and identity. Stones could become bread, a mountain leap provoke an angel catch, and all the world's empires lie at Jesus' command, if . . . Each temptation is refuted with reference to a verse from Deuteronomy when Moses elaborated on the conditions of the covenant and how the people were promised a bounty of agricultural and urban well-being *if* they kept the covenant (Deut 8:7–18). As with Moses, the lessons learned in the wilderness were to be the guarantees of the blessing of God, for human collective well-being.

Another author, later named Luke, claims he offers a "well-ordered" account, with a Gospel from "eyewitnesses and servants of the word" (Luke 1:3). He takes the same wilderness temptations, but how does anyone know it happened, if it was only Jesus who was there? The order of the second and third temptations is changed, and emphasis is on the role of the Holy Spirit: "Jesus full of the Holy Spirit . . . was led by the Spirit . . . and returned in the power of the Spirit into Galilee" (Luke 4:1–13). This is the story of the Spirit of Jesus being let loose on the world, and surprisingly for a Gospel with more than average reports of angels, none are mentioned in this wilderness experience, nor wild beasts. Instead, time is key. It is a holy time, and Satan leaves Jesus when his temptations fail. The opportune time would come: "Then Satan entered into Judas called Iscariot . . . he went away and conferred with the chief priests and captains how he might betray him to them" (Luke 22:3).

These writers had no concept of a Green reading of their story, but they may have given us a profound insight into the spiritual life of Jesus of Nazareth: solitary, mystic, nourished, and inspired for his life's goal, through this wilderness immersion. It was this experience that formed Jesus' resistance and resilience to see things through. It is this Green Jesus who would go to the cross.

John's Gospel was written later than the Synoptics, with different content and style. We find no wilderness experience nor interest in how Jesus developed his sense of identity. He comes as a given, with an altogether new level of reflection, and predates creation itself: "In the beginning was the Word" (John 1:1). This is a cosmic vision of Jesus of Nazareth, but echoes of a wilderness experience remain, nevertheless: the repetition of the story of the Jews who ate manna in the wilderness, with the sign of the feeding of the five thousand, and Jesus' claim to be the bread of life. The majority of this Gospel is concentrated on Jesus' last days in Jerusalem, and it records his retreat to the hills, to escape crowds and rest (John 6:3, 15); nature is his resource.

The Synoptic Gospels emphasize Jesus' faith identity was formed through the austerity of the wilderness, and yet Jesus repeatedly takes nature as being inherently fruitful, such as the poor fig tree, taken to symbolize Hebrew faith practice and cursed for fruitlessness (Mark 11:13–14). Nature provides, as an expression of divine power; wilderness and providence are intertwined, as an oasis in the desert. Then "look at the birds of the air . . . consider the lilies of the fields" (Matt 6:26, 28). Nature is the basis for hope and courage. Providence is used to enforce one of Jesus' most difficult teachings, to love your enemies: "But I say to you, love your enemies and pray for those who persecute you so that you may be children of your Father in heaven, for he makes his sun rise on the evil and on the good, and sends rains on the righteous and on the unrighteous" (Matt 5:44–45).

How much of this goes back to Jesus himself, we can never be certain of, but it makes sense for a teacher who had a faith formation in the wilderness to share a strong awareness of divine providence. It encourages us to dialog with Jesus when we are in wild places ourselves. Jesus knew life from the ground up, from survival and dependency on the land.

PROPHECY AND PARABLES

The stories of Jesus emphasize his public ministry began through an experience of wilderness immersion. It shaped his prophetic role in how he spoke and behaved. His time alone with angels, wild beasts, and the devil was his springboard for action and self-understanding (Mark 1:13).

Jesus knew the injustice and violence of Roman "peace." Pax Romana meant a subsistence life for his people, with unemployment and hardships that drove some to violence. Jewish yearning for better times had become a desperate hope for a Messiah to bring more radical change. Some took things into their own hands for survival as bandits, fraudsters, tax collectors, and prostitutes. Instead, Jesus proclaimed the coming realm of God with who, how, when, and where he healed, ate, and slept. These memories have come through inevitable overlays of the church, with its needs for miracles and glory. Yet there are parts of the Gospels where we are more likely to find his mindset and message: in his parables. They have ways of working in our minds that make the details stick. If we give them attention, they can still disturb the status quo by being indirect; we are left hanging. It seems Jesus had a genius for choosing characters, stories, and punch lines that embedded in peoples' minds.

Parables are not just a Jesus thing. Robert Short offers a variety of descriptions: they are a means of speaking truth to power with survival in mind, and they have a recognizable structure. If Picasso said, "Art is a lie that makes us realize the truth," some let the teller get away with murder. Jesus' parables may be art parables as they often address powerful men, just as Shakespeare knew: "The play's the thing / wherein I'll catch the conscience of the King."[12]

The contradictions to conventional expectations in the outcomes of many parables, such as the Samaritan traveler who helps his beaten Jewish neighbor, or the late vineyard workers who are generously paid, carry the sense of "you couldn't make this up . . ." These are different than the recollection of Jesus' life. Once told, who would feel they could nuance a parable, when it has that sense of careful and pointed construction? There can be a more direct route back to Jesus in the memory of his parables than in any of the events around them.

Parables leave us with questions. They draw us in with features that have some grit and with a strange realism, because they are so obviously contrived. These resonate with Hebrew faith history but have a freshness, or imprint, that arguably delivers Jesus' approach.

---

12. Short, *Parables of Peanuts*, 12; citing Shakespeare, *Hamlet*, act 2, scene 2.

For no teacher would be credible without reference to the past, nor would Jesus' faith make any sense without a profound knowledge of his peoples' history. It predisposes Jesus to tell parables. The ancestral myths of Jewish origins describe human weakness, even with heroes of the faith like Moses. My favorites are the patriarchs and matriarchs of Genesis. Give yourself an evening to read from chapter 12: Abraham and Sarah laughed at God; Isaac did not recognize one son from another (or did he?). Jacob the cheater gets cheated himself, by Uncle Laban. Did Jesus have Jacob's decision to go home in mind, with his story of the prodigal son and forgiving father (Gen 32; Luke 15:11–32)? Then, consider the tumultuous period of the first Jewish kings, when the glorious King David turned into a self-serving letch and abused his power to have sex with a married woman, Bathsheba (2 Sam 11:1–5).

These stories admit human beings are complex and that Jewish faith history was full of the consequences of failure, as well as triumphs. It laid out the role of prophets who told parables. A cryptic message was a wise strategy to avoid the violent reactions of powerful men and occasionally women, like Jezebel who troubled Elijah. There are many in the Old Testament: the eagles and the vine (Ezek 17:2–10); trees making a king (Judg 9:8–15); the wasted vineyard (Isa 5:1–7); Samson's riddle: strong bringing sweetness (Judg 14:14); Nathan and the poor man's ewe lamb (2 Sam 12:1–4).

Jesus of Nazareth grew up with these parables and how they were used to challenge power. When he developed his own, he drew on his prophetic nature, even his genetics!

Matthew begins his Gospel with a genealogy of Jesus, to show he was a son of David. This was a feature of the Messiah, the Christ. But King David could trace his origins to Ruth, a Moabite woman who eventually married into Jewish line through loyalty (or love) for her Jewish friend Naomi (Matt 1:13; Ruth 1:22). So not such a surprise that Jesus placed an outsider, a Samaritan, as a hero to illustrate the greatest commandment: to "love your neighbor as yourself" (Luke 10:25–37). Jesus told "the good Samaritan" parable to point out a fault line in Jewish life of those who thought themselves good Jews. It was the outsider, the faithless Jew, who fulfilled Torah. He was not the person to invite to public occasions unless you liked drama.

Jesus taught during times that Jewish identity had lost its way. In the best of times, the mindset of being a chosen nation, which messed up their side of the bargain, made Jewish believing complicated; add on successive invasions by superpowers, exile, and loss of national independence. Rome was the latest occupant and took away the sense of abundance that was core to the promises of God. Contradictions to claims of belief were found everywhere and drove faithful reflection below the surface of how things are.

THE GREEN JESUS?

Motives, the inner life, and freedoms that allow people to behave in kind or unkind ways—these became the primary resources for a teacher who would not abandon the practical consequences of life. They explain how parables resonate even for an integral worldview today.[13]

The paradoxes of parables were extraordinary as a teaching strategy. Rather than painting a description of another world, Jesus taught God was present in language that was adjacent to reality. The realm of God was close, and parables his chosen means of communication.

This sort of a teacher inspires deep reflection, including self-criticism: am I as much a part of the problem as my enemy? This became a crisis for his followers as Jesus went on to act out his own agonizing parable. His inner circle unraveled, with betrayal, abandonment, and denial after his arrest. The power of the church came through this terrible experience to discover Jesus was right. To have met the Messiah and have failed utterly, but still be forgiven and loved, provoked devotion and praise. Jesus' memory-making stories were treasured.

Matthew's Gospel takes God's realm as the subject for most of Jesus' parables.[14] It is compared to a field sown with good seed, but then with weeds that are inseparable until harvest (Matt 13:24–30); like a mustard seed that becomes a tree: "so that the birds of the air come and make nests in its branches" (Matt 13:31–32); yeast mixed in with flour (Matt 13:33); treasure hidden in a field (Matt 13:44); a merchant in search of fine pearls (Matt 13:45); a net thrown into the sea (Matt 13:47), so to know this is like the master of a household who brings out of his treasure what is new and what is old (Matt 13:52–53). There are prophetic warnings of judgment in these teachings, but all point to positive outcomes if you pay attention. The created order illustrates the realm of God and calls for human disorder to be resolved: thy will be done.

The references to nature in these parables make a case for the faint Green Jesus. But I am unconvinced. It was natural to use nature, that is all.

The parables that show Jesus had a focus on the everyday struggle to get by are more important. Jesus noticed money, household life, and human yearning. These themes will become critical for the gospel of Jesus Green:

> Lazarus is ignored at the gate of the rich man. (Luke 16:19–31)

---

13. Wink, *Engaging the Powers*, 5–6.

14. The Greek word usually translated as "kingdom" is *basilea*. *Basilea* implies no gender, as does the phrase "realm of God," rather than "kingdom of God." So, I prefer realm. Monarchies are not as they were, but there are definite "realms." What of the realms of Facebook or Windows; music, fashion, or sports?

The rich (young) man cannot give up his wealth; in fact it's easier for a camel to pass through the eye of a needle. (Matt 19:24)
The dishonest steward is praised for his shrewdness. (Luke 16:1–13)
The widow gives the little she has to the temple. (Mark 12:41–44)
The traveler falls among thieves. (Luke 10:25–37: the good Samaritan)
The younger son spends his inheritance in reckless living. (Luke 15:11–32: the prodigal son)
The woman rejoices to find a lost coin. (Luke 15:8–10)
Three men are given talents of their master to make the best of while he is gone. (Matt 25:14–30)
Vineyard workers get equal pay. (Matt 20:1–16)
The real owner of the vineyard is denied rent, with dire consequences. (Matt 21:33–41)
The wedding banquet is delayed, and the waiting divides bridesmaids. Weddings promised owners continuity and transition. (Matt 25:1–13)

These parables acknowledged the bigger picture in terms that were very familiar. They made memories in the way they questioned the status quo or validated ordinary life as special. Some scholars highlight how the parables have parallels in earlier foreign traditions, but the Gospel writers gave them to Jesus with these themes, and Jesus himself may have collected sayings.[15]

Most of us have lived through financial crises, but the Roman occupation made money a chronic problem. Taxes that took wealth to Rome as tributes pressurized the whole society, and there was one location where it was felt most deeply. The Jerusalem temple represented national life even under occupation. A riot broke out against Pilate, before he met Jesus of Nazareth as king of the Jews, because temple treasure had been sold to build an aqueduct.[16] Such Roman atrocities are understandably not mentioned in the Gospels, and Pontius Pilate is portrayed as a reluctant executioner.[17] It is important to appreciate Jesus' parables played off historical and social realities that gave them more bite. For critical voices like Jesus', the temple was much more than the meeting place of failed religious leaders: it was where this failure was practiced on the poor majority. Jesus foresaw its destruction

---

15. Funk et al., *Five Gospels*, 106.

16. Theissen, *Shadow of the Galilean*, 3–5; *Wars*, 2.175–77, in Josephus, *Complete Works*, 730–31.

17. Horsley, *Jesus and Empire*, 33–34.

through his own death: "I will destroy this temple that is made with hands, and in three days I will build another, not made with hands" (Mark 14:58).

When Jesus decided to act against the temple, to turn over the tables of money changers, all his parables about money became more significant. With imagination we can take this prophetic, parabolic power of Jesus further. If we can recognize our role in present-day systems, surely Jesus did for his own. Who am I to criticize today's elite for destroying rainforests through multinational systems, if I am not making changes myself?

Yet imagination is needed. We cannot reach through the faith of the Gospels to claim much about Jesus in history. This fragmental Jesus must always be faint Green. But I love parables. They are fascinating and humbling in a good way when I take them seriously. Is it possible that parables, and thinking through parables, nurtures a sort of self-awareness and empathy? Can this contribute to collective change today, for a tipping point of national and international relations? It may ask too much from parables because they are deliberately so open to interpretation. But goals based on the wisdom of parables may be part of the transformation of our systems and governance to conform to our better human nature and the providence of the earth. There is an existential parallel between prayer and parables that arguably comes from Jesus. Parables often describe God's realm, and in the Lord's Prayer Jesus suggested we pray three times for the realm of God: to come, to be done, and to be done on earth as in heaven. It leads to a request for daily bread. How more practical can this be? To move from parables to actions was something I presume Jesus intended. You cannot determine outcomes, but it sets a process.

How would this prophetic, parable-telling spirit of Jesus be alive today? It is not difficult to imagine Jesus would highlight the relationship between poverty, money, and the environment; many poor communities are the first to experience consequences of mining pollution in water supply or deforestation of their own lands with flooding and crop failures. This is a faithful representation of a Green Jesus for our times because his prophetic spirit has reached us, despite all the twists and turns of his identity over the centuries.

I can almost hear Jesus telling a parable about two farmers, one poorer and one richer; the poorer could not afford fertilizer but knew how to get by with land that lay fallow, versus the richer whose crops were so abundant they left no ground to spare: when warfare comes and fertilizer runs out, who is the wiser? Whose sons and daughters will be able to live off the land?

I have surprised myself in digging deeper into what I believed would be a limited exercise on the historical Jesus being a Green Jesus. Granted the hesitations of any facts claimed about Jesus, the presence of Green insights in the little I found is important. Some of this is because our modern and postmodern experience is bizarrely cut off from nature; so many ancient teachers would strike us as Green. But Jesus of Nazareth had an interest in money and social order and drew on nature to make his points. These memories of Jesus offer bridges between the economy and the environment, in parabolic forms.

A remote and uncertain Jesus is frustrating when we want clearer guidance for our crisis of species loss and climate change. Yet there is a message: money matters, and simple answers are suspect. If a rich man *can* get into heaven, a fight against financial exploitation cannot be reduced to single answers or guilty groups. This is a Jesus thing: to drive for understanding. I would say this is biological, and his parables nurture a deeper Green.

Chapter 4

Deeper Green

SERIOUSLY PERSONAL

I felt my eyes growing. Did I really see that? As a questioning teenager I was not close to my father, but he gave me a child's experience of wonder that changed my life. Our summer holidays in Devon and Cornwall delighted me with beaches and ice creams. Then Dad showed me the mystery of the low tide and how to be still, to wait and watch.

I fell in love with the life that teemed in rock pools and the power of the Atlantic Ocean to reveal new things each day. These marginal worlds teach deeply. I improved my agility to leap across rocks and learned patience and humility to accept rhythms in tides. Despite being an extrovert, I liked to keep this exercise to myself, jealous of finding the best pools, as if the creatures that emerged were a private treasure. I knew I would never fully comprehend these worlds, as the rock pool is connected by the tides to the abyss of the oceans. There was always more, and that was what drew me.

As children we have played in the tension of knowing and not knowing, imagined horrors and beastliness, for the thrill of being scared. Then the teenage recklessness to jump into water, not knowing its depth, for the rush of adrenaline and release from panic, is practically animal. With sadness we learned the shadow of death, losing a pet, a grandparent, or worse. The experience of loss, sooner or later, broke a carefree paradise. Our personal garden of Eden was spoiled, if we ever walked there.

When I wanted to observe a rock pool I would kill a limpet, the conical mollusk that clings to the rock with a big single foot. A little violence to entice the life I found captivating, but I did think occasionally about the limpet. Death as part of life.

I was not being original. In 1963, the year after I was born, zoology professor Dr. Robert Paine chose Pacific Coast rock pools for a landmark experiment in the history of ecology.[1] He simply removed the top predator, the starfish, and waited to see the impact. Sea urchins multiplied and ate or smothered the algae, barnacles, and seaweeds they shared space with, and biodiversity collapsed. Such a clear demonstration that predation was vital to the health of those rock pools raised the importance of predator-prey relationships for ecology in general.

Jesus may never have seen tidal rock pools, but he knew Galilee and lakeshores that could teem with life and asked, how could this be? Jesus reflected on what made abundant life possible but saw how the issues of life and death in nature played out in Roman occupation of his country. In the truth of predator and prey relations, life was uncertain, so what could be relied upon? Jesus rejected the cycle of violence of an armed rebellion and encouraged a search for deeper truth. He shared a parable about two men, one who built a house on sand versus one who built on a rock (Matt 7:26–27). I have never built a house, but rock is fixed and uneven. It challenges a builder. So, the parable challenges us: What is there to build life on? What is this rock? Is life building a house? Even if you had been with Jesus, you would still be asking questions.

Just like I found the drama of the creatures in rock pools unforgettable, the stories Jesus told made memorable impacts in the mindsets of millions. It is easy to picture him by the lakeshore teaching about houses, rock, and sand. Did he embody answers to the parable? The rock is to be centered on God. To build a solid house was possible, despite foreign domination and the rampant corruption of people caught up in it. Jesus and the house will become a paradigm for Jesus Green, as Jesus set his face to Jerusalem and the house of his Papa, his personal God: a bad end for an extraordinary beginning.

When I think of Jesus' story and rock pools, they share an obvious truth: if you do not go beneath the surface, you miss the life they hold. Granted, sometimes I saw surprising action on the surface of the pool, like a crab eyeing my rude arrival as she ate her meal, it was the underwater world I explored, just as Jesus taught under the skin of being human.

Let's stay with water, to accept some facts. Much of the adult world is false: we take the boundaries created by employment, routine, and family expectations as fixed and unquestionable when they are just the surface of agreements; whereas open-endedness, consciousness of inner and outer realities, and the connections between them are what make things *deep*. I have an inner reality that perceives an outer world. Shake off how much of

---

1. Carroll, "Ecologist Who Threw Starfish."

your day is familiar to you, and be that child gazing into a rock pool, not knowing what will happen.

Immediately I am caught up in a process. It is more than rock pools and ecology; it is especially about what we do as human beings in response to the foundations of life. I have led us into our childhoods and self-evident changes in our worldviews. Deeper Green asks, are we growing again? Just as when we were children, is our consciousness changing, despite us, because the truth of the beyond demands it?

You know this experience because it comes with tell-tale anxieties of recognizing you are not in control. There is a sense of alienation from people who are not on the same wavelength and from loss of securities for the future. Deeper understanding can increase a sense of being powerless because we realize the changes that are required are deeper too.

A little like my father led me to appreciate rock pools, the mothers and fathers of a deeper Green gave foundations we can build on. They describe how we belong to living systems and are wise to build on the sense of how you and I are part of the whole.

There is a story from the 1960s: *A rich man sees a poor man lying in the sun all day doing nothing but fishing with a line. He shouts out, "Hey there, why don't you get yourself ten lines, then buy a boat with a net, so you can get ten boats and send the others to fish, and have a good life?" "Why's that?" asked the poor man, "So that I can lie in the sun all day doing nothing?"*

This story fits a classic early Christian Green book, *Enough Is Enough*. "Enough" for what? What for me is enough? Well, that depends on what makes me happy. But is that it? Is my happiness the goal of my life? Or is it part of a bigger story? What is enough for *us*? And who can or should answer that one? Now I am collecting questions like a hot barboy gets tips! Environmentalism leaps to philosophy. It must chase out the fake top predator of our minds, the surface that confuses needs with desires. I want to insist on this movement. Deeper Green is this shared consciousness of belonging to an extraordinary planet, of staggering complexities and simplicities intermingled; me and we; a growing, felt "Wow!" of existence that has often been restricted to religious mystics, but now has universal appeal.

In chapter 2, "What Do You Mean, Green?," I highlighted the watershed book *Silent Spring* by Rachel Carson as a mark of the start of the modern environmental movement.[2] The book goes deeper because its warning came from understanding life processes, not just outcomes. I was delighted to discover Carson first studied and wrote about the life of the oceans.[3] *Silent*

2. See ch. 2, p. 46.
3. See https://www.rachelcarson.org.

Spring proved change was possible through public reaction. The science of how substances flow through a food chain led Carson to sound the alarm about pesticides, like DDT, that had decimated hawk and eagle populations. Through her work and the response to it, natural and human systems were shown to be related. There is a contemporary struggle to protect the pollinators who are killed by insecticides based on nicotine or glyphosate.[4]

The Systems View of Life: A Unifying Vision, by Fritjof Capra and Pier Luigi Luisi, is a compendium book with fabulous breadth, which places environmental insights in a bigger picture. It offers a springboard to participate in hopeful change in the way it joins different subjects by common principles. Reading it brings the *experience* of knowledge being more than the sum of its parts. Like learning to read, these insights liberate a sense of being alive. If this is too academic for your taste, there are books that explore deeper Green through personal journeys.[5]

As I discovered this thinking and the way it demands change, I had some "déjà vu" of Jesus as a teacher with the awareness of systems, not just individuals. When Jesus healed, he knew it had wider implications. The living world is better understood through ecology, but a deeper Green approach demands this knowledge is brought to human endeavors. It comes back to us, our systems; it's personal. You and I are part of the problem *and* the solution.

My personal deep Green awareness coincided with how gay culture entered public headlines in the eighties. I was invested in both. Both were mocked. It was a gay plague in San Francisco, soon to be named AIDS, and it was laughable "tree huggers" with weird descriptions of Earthmother taking her revenge on industrialists—those "deep Greens." Now I realize this timing was not a coincidence because both are about human civilization reaching for maturity, to understand our sexual diversity (a form of inner life) and our place in the biosphere (the outer realities). Deep Green awareness rises as the environmental and species extinction crisis worsens. It parallels the way LGBTQ+ people are being integrated into the mainstream, or not.

Silent Spring was key to the insights of the Norwegian philosopher Arne Naess (1912–2009), one of the founders of *deep ecology*. Naess was also marked by seeing rock pools! He worked with American environmentalist George Sessions (1938–2016) to propose eight basic principles to deep ecology, such as a decrease in the human population if nonhuman

---

4. Dengler, "Neonicotinoid Pesticides."

5. E.g., Abram, *Becoming Animal*; McCarthy, *Moth Snowstorm*; Kimmerer, *Braiding Sweetgrass*.

life was to flourish. Already human interference in nature was excessive and accelerating.[6]

Today's mass species extinctions and rapid climate change back up these insights. Humanity has had such a marked negative impact on the biosphere that geologists have given our era the name Anthropocene. We are back to building life upon sands, and there are plenty of capable people to reassure you, "Do not worry," when now is the time to be alarmed—a message that is a contemporary form of wickedness.[7]

The insights of deep ecology highlight what must change.

The number of human beings alive today is roughly eight times the number alive in 1800.[8] We expect to live longer and consume more than our forebears. The United Nations has recognized the potential for disasters due to this human load but failed to agree on strategic change. The solutions are fraught with difficulties such as how freedom of choice for women coupled with improved basic conditions for households can mean reduced family size, while resources for this fall short of the targets for change.[9]

Human overpopulation still comes with a staggering degree of mass poverty. I do not find the predictions of higher but stable population levels later this century morally OK, not when this means vast shantytowns and climate refugees. It fails to grasp the risk of collapse of whole living systems with the pressure of human impacts. This deeper Green reality is seriously personal. The extreme Chinese policy (1980–2016) of one-child families produced its own challenges and came with violence.[10] There are other solutions, but they will not be explored or negotiated without shifts in public opinion to face the crisis.

One overused word fits this global drama, which demands human genius and courage: *unprecedented!*

The seventh principle of deep ecology from Naess and Sessions indicates where solutions lie: a different orientation, to appreciate quality of life over increases to personal income. This is a preference for "dwelling in situations of inherent value."[11] It carries a sense of home.

There is content to how "enough is enough," how conversations on family size can take place humanely, based on the common values of quality of life. It raises up the value of women's welfare movements as allies

---

6. Devall and Sessions, *Deep Ecology*.
7. Tupey and Pooley, *Superabundance*.
8. Roser and Ritchie, "World Population Growth Changed."
9. UNFPA, *Strategy for Family Planning*, 7.
10. Wang and Zhang, *One Child Nation*.
11. Devall and Sessions, *Deep Ecology*, 68.

for deeper Green work. This will not be by force but by a shift that we can already share in, to find enough is enough. The house can be built on rock.

It also roots us in that truth we knew as children of sharing and belonging, in families, in homes, in nature. The child's love to explore in nature or in a computer is without prejudice. It is all good. The sense of connection you had, being content in play, reassured by a parent sending you to sleep, familiar sounds and smells of home and garden—these things speak of a knowledge of what is good. To be denied this is to carry pain in life, because they are fundamentals.

The sense of home is timeless, and Jesus knew it too, because he took a child to teach his core group what matters. Unless you become like a child, he taught, you cannot receive what I am teaching about the realm of God (Matt 18:2–4), and when his rivals asked him about the future of this realm of God, Jesus refused a time line. Instead, he relocated it to the inner life, like the child's view that connects with everything, without prejudice; the realm of God is within each of us (Luke 17:21).

This destroys the tyranny of finance and scarcity we have learned as adults and welcomes "enough is enough." It welcomes complexity. There are so many things, so of course things are complicated, and isn't that extraordinary and essential? Then populist, simplistic messages for big problems are exposed as manipulative and disrespectful of the glory of life.

The Green of Jesus Green searches us out in a personal way and calls us to change. We have a future living by facts of rock, lakes, and oceans, not by dollars alone. To recover this trust is the prime human endeavor of this century, and it promises more than enough. This is to respect the wonder we have experienced as children and adults for the living world and the truth of our part in living systems. It is this seriously personal dimension to deeper Green that promises sustainable change because it comes with the profound satisfaction of growth in the human spirit.

THE POWER OF OIKOS

Our home in Montreal is full of plants, especially stunning miniature species of codex and succulents. Some are active in winter, so I come upon a plant in flower or new leaf appearing as if by magic. We can notice patterns to this process of emergence, and its not just for plants.

I had to be given a long job title for me to notice the power of a short Greek word, *oikos*. It was in my head and I never noticed. Millions know it as a brand name for yogurt, but it means "house" or "household."

What I discovered came through a tale of three university cities: Liverpool, Cambridge (UK), and Montreal. Each contributed to my heightened sense of home. But not in the way you may expect. First at Liverpool, I was a naïve student enthralled by human knowledge, life, and choices. I studied marine biology. Rock pools again! It was also a "coming out" time, as I mixed science and spirituality, travel and falling in love. I juggled intense personal experiences with engagement in the church and what science meant to me as a career. It was a unique mix, but I was always joining the dots.

I learned that at every level, life relies on interdependency. The life of cell membranes, gates, and mitochondria resonated with the dynamics of seashore ecosystems. But it was the wider life of community and the church that took me forward.

In 1981, I was chosen to be in a young adult team for a six-week trip to the Methodist Church in Jamaica and the Bahamas. I was nineteen. This was a trial for a youth exchange program and my first flight and first trip outside Europe. It made me say "amazing" too often. You know that tropical heat, turquoise water, and white sand beaches bring out feelings of freedom and desire. Throw in an athletic boy with a charismatic voice, and it was time to name things, in the safety of strangers far away from home.

Our trip was judged a success, and a national youth work job was created. The timing matched my graduation year, so I applied and was offered the job, as proof of the process. This national and international youth work based in London led me to become a Methodist minister, with training through the illustrious Cambridge University.

We can find the best and the worst of things in a university, just like the wider world. They can be places of oppression or uprising. I read theology, which was supposed be the "queen" of knowledge and span the great divide of the arts and sciences. There was logic to the arrogance of Christendom: theology claims universal truth, and universal truth is sought in universities. I love theology, but this alerted me to question any department as authoritative alone. Thankfully, as theology has moved from "premier league" to the "minors," there has been a recognition of the role of interdisciplinary studies. I first witnessed the strength of this at Cambridge, through a war studies seminar and the presentation of anthropology, history, sociology, economics, and theology in a single day.

Twenty years later, I returned to university life in Montreal, with the job title that focused my attention, McGill Protestant *ecumenical* chaplain. I worked for two denominations, The Anglican Church and The United Church of Canada. *Ecumenical* meant cooperation between denominations. As staff, not student, I had a new perspective, with time to notice the underside of things. Two decades have seen greater economic pressures on learning, as well as weaker churches. I found contradictions commonplace and echoed in other universities, both sides of the Atlantic. The governance of the university was inconsistent with the knowledge its professors were teaching. Silo thinking was still normal and interdisciplinary studies marginal. Tensions regularly surfaced between a board of fundraising governors and a senate of educators. Student Services, to which I belonged, was underfunded and overwhelmed by rising student needs.

As ecumenical chaplain, I decided to be different than my predecessors and relate to the general spiritual needs of students. To offset popular cynicism that studying was just learning to pass exams, I invited people to join me on a walk round about the campus, to question our goals: "What is education for?"[12] We would stop at several points to take time to notice the surroundings and share our discoveries, like the Natural History Museum, a library entrance, tree stumps on the adjacent Mount Royal, and a starry ceiling in the original university building. As participants bounced ideas off each other and the surroundings, it was natural to question their education, and life itself. I was doing my job!

I called this work the Oikos Project, because it offered interdisciplinary perspectives and I knew *oikos* as the Greek root word for three other words. Two are familiar, and obvious for this book: ecology (*oikology*) and economics (*oikonomos*), but the third is less well known and was the gift of my job title: *ecumenical* (*oikoumene*). For the exercises to make more from the stopping-off points I drew on the insights of Joanna Macy and Molly Young Brown.[13] Oikos walks helped to refresh students' energy, sense of self, and vocation—to be "at home." Home was emerging for me, not as a place I lived in but as life itself. On each walk we realized how I cannot be "me" without "you," without "us," and that the "us" is huge.

I cannot take you on an oikos walk, but I can remind you of how such holistic insights have been offered on the page.

Below is a familiar graphic, a Venn diagram. British mathematician John Venn first conceived it to explore and analyze sets of numbers. But its general use was appreciated straightaway. I like this example because it teaches how to be more resilient, as life throws us curve balls.

---

12. A question asked by Orr, *Earth in Mind*, 131.
13. Macy and Brown, *Coming Back to Life*.

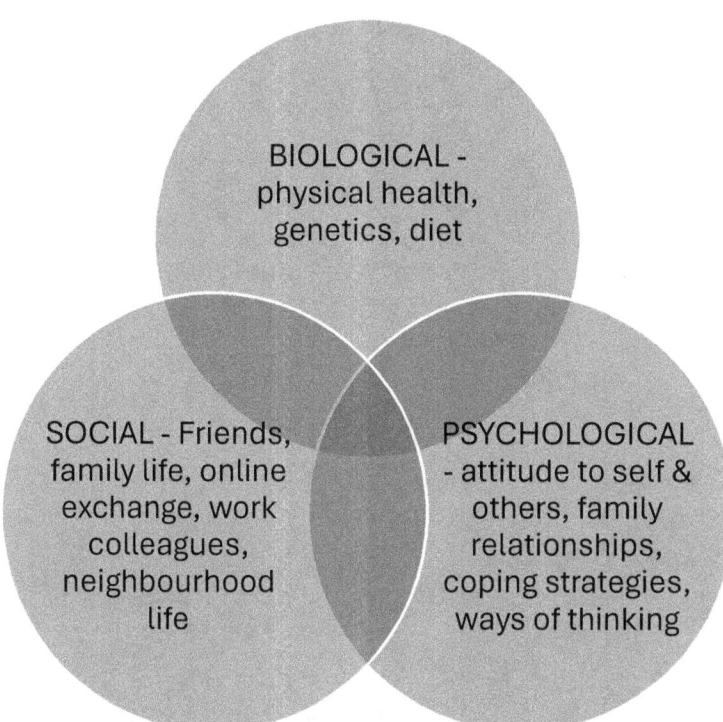

A Biopsychosocial Model of Health[14]

Health has a physical, mental, and social dimension. We can add a fourth, spiritual health: the sense of meaning, connection, freedom, and joy. Each is an inseparable aspect of the whole.

If we are weaker in one sphere, we draw on the strength of the others to find well-being: if I lost my job, I may be understandably depressed (a mental health need) and I would lose friends (social health down), but aerobic exercise can offset depression and doing it in a team sport may be even better. An activity that is found in overlapping areas, or especially at the center, is arguably the most effective. Try a drumming band?

---

14. Adapted from Seth Falco, "Biopsychosocial Model of Health 1," Oct. 26, 2016, at https://commons.wikimedia.org/w/index.php?title=File:Biopsychosocial_Model_of_Health_1.png.

For Green concerns, the Venn diagram has already been used to help understand the complex challenge of sustainability.

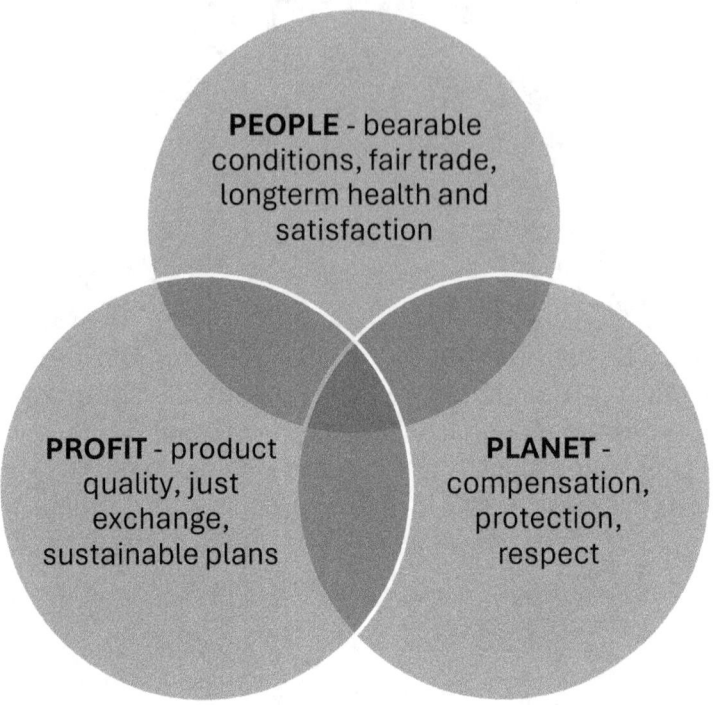

People, Profit, Planet: a recognition of the necessities of each sphere points to sustainability at their coincidence and brings a less conflictual perception of life.[15]

15. Adapted from Arowoshegbe et al., "Figure 1 People Profit Planet," 2016, at https://commons.wikimedia.org/w/index.php?search=Arowoshegbe&title=Special:MediaSearch&go=Go&type=image.

The Oikos Venn diagram is simpler, with the deeper association of a common root word:

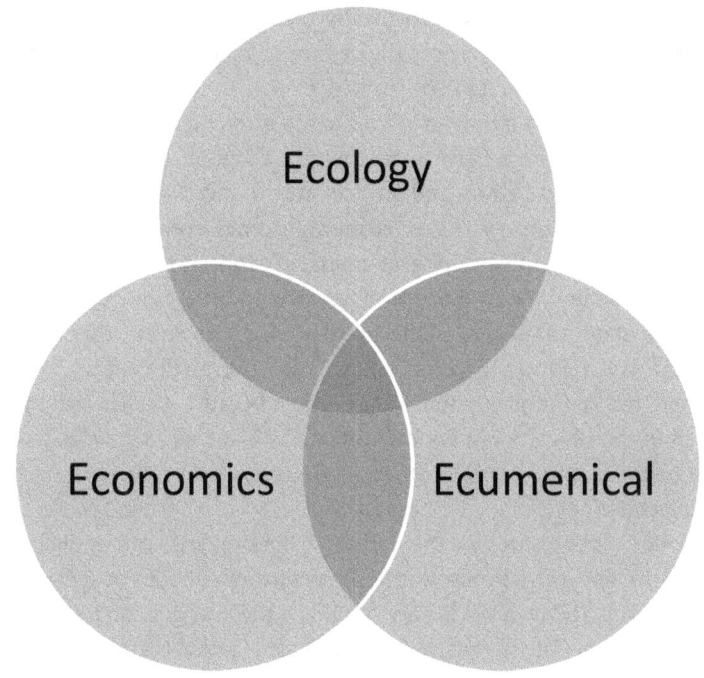

Oikos Venn diagram

We all know home and the household are vital. "Are you at home?" has unique answers for you and for any group you belong to. It is not something someone else can or should take control of. To come home, to talk about your homeland, or to know your ancestors' home resonates deeply.

Grasp these three words and how they overlap and you are on the way to becoming eco-literate and deeper Green, at least in your understandings. It is easy to remember and to share.

> *Ecology*: The life of the household
> *Economics*: The management of the household
> *Ecumenical*: The relationships within the household

This Venn diagram is a true presentation of how we are embedded in the home. The life of the household could be overmanaged and is undoubtedly better if relationships in the household are good. Management of the household will go astray without knowing about the life of it or the relationships within it. Relationships in the household will soon become strained if it is badly managed or someone in the household is sick.

This oikos trinity helps us to increase our trust in holistic wisdom, just as Jesus said humans do not live by bread alone. It is no coincidence that oikos has been chosen over the centuries for describing fundamental human and biological activities:

Economics has a strong grip on factors in our daily life, but it is not a science. Despite impressive mathematics, economics is a "best-guess" understanding of how goods and services are produced and exchanged. The *Oxford English Dictionary* tells us "economy" comes from Greek *oikonomia*, with the verb *nemein*, meaning to manage. Modern use dates from the seventeenth century with the rise of science and the Reformation, when the nation-state was taken as the household of the ruler. This "political economy" was replaced by economics in the twentieth century.[16] Of the three spheres, economics stands out, not just because it takes a dominant place in human society but because often it is taken as a reality in itself. It would be the first to say to the others, "I don't need you." This history helps us to question.

Ecology is the study of the nature of life on earth and explicitly the interactions between species and their environments: it recognizes how different forms and levels of life are related by the flow of matter and energy in living systems. The German biologist Ernst Haeckel is credited with the term (1866). The shared etymology of ecology and economics was anticipated by botanist Carl Linnaeus, who wrote of the "economy of nature" in the eighteenth century. Charles Darwin admired Linnaeus' work and used it in his *Origin of Species* as "the economy or polity of nature."[17]

Ecumenical has a rich history that is lost to many modern dictionaries. There are *seven* different meanings to the word, spanning centuries, that I have simplified to three.[18]

Every Christmas we can hear one of the earliest uses of ecumenical, and it is already linked to money! Luke dates the birth of Jesus by the order of Caesar Augustus, when Quirinius was governor of Syria, that the "whole world" be registered (Luke 2:1). Mary and Joseph are forced to travel to Bethlehem. *Oikoumene* is the word translated to "whole world." It is found in the logo of the World Council of Churches (WCC), where *oikoumene* means "the *whole inhabited earth* . . . it reflected the interaction of religion, philosophy and political administration as they shaped society. In modern

---

16. Capra and Luigi, *Systems View of Life*, 48–52.
17. Kormandy, "Ecology/Economy of Nature," 1292.
18. Visser 't Hooft, "Word 'Ecumenical.'"

usage, the word embraces the unity of God's whole creation and recognizes every human pursuit as subject to the healing ministry of Christ's Spirit."[19]

The WCC definition also reflects how ecumenical then became an "in-house" word for *the whole church*, in its attempt to reach universal understanding. If you find the Trinitarian doctrines of God the Father, Son, and Holy Spirit meaningful and true, then you have benefited from the ecumenical councils of the church as "having universal appeal or validity." These fourth- and fifth-century councils established consistency and boundaries of belief as Christianity was adopted across the Roman Empire.

A third nuance for ecumenical came out of the later schisms of the church, in reaction to chronic bigotry and hatred. Regular churchgoers have heard of the ecumenical movement dating from the early twentieth century, *as the effort to reconcile differences between Christian denominations*. The desire for unity between churches has gripped some of the finest church leaders, not least because divisions were hinted at in the second century by the Gospel of John, when Jesus prayed for his followers to be one (John 17:21).

Although ecumenical is the lesser known of the oikos trinity, it has demanded courage and inspired a healing vision: think of the impact of being ecumenical in Northern Ireland and now in Ukraine. It is also the oldest of our three words and in many ways the key one, since the question of how human beings *inhabit* the planet is *the* question of being Green.

Oikoumene has a spiritual dimension from its emphasis on relationships, and despite the controversies, I can still get frissons of this at the ceremonies of the Olympic Games. We are one species, and to bring together nations, races, cultures, and ethnicities is something bigger than sport. Billions of locations are connected "live" worldwide, as a global household: Homo sapiens can look in the mirror. Ecumenical has a technological boost.

Three vital dimensions have been brought together because the home has a vital reality, beyond human life. The Oikos Venn diagram is a value-neutral tool to conceive solutions and expose failures.

Take Canadian nuclear energy policies that lack coherent federal standards. A surface-based nuclear waste dump, termed a "near-surface disposal facility" (NSDF), has been proposed in southeast Ontario. But this is within the watershed of the Chalk River, a tributary to the St. Lawrence River. The environmental organization Sierra Club noted the dump will include particles normally rated intermediate-level waste (ILW).[20] These

---

19. World Council of Churches, "About the WCC Logo," para. 2.

20. Sierra Club Canada Foundation, "Oral Presentation," 2–3. The Sierra Club was founded in 1892 as a walking club by John Muir et al. (https://www.sierraclub.ca).

require heavy lead covering, so how can a near-surface disposal facility be acceptable? ("ILW generally requires a higher level of containment and isolation than can be provided in near-surface repositories."[21]) Impacts of any leaks will flow into the St. Lawrence River and touch millions of living beings, including humans, downstream.[22]

This is a classic case of economic thinking proceeding without ecumenical or ecological input. The tsunami that hit Fukushima in Japan in 2011, and the war in Ukraine that disrupted the Chernobyl nuclear power plant, illustrate how plans must anticipate major disruptive events, which will sooner or later take place. For such a long-term danger, ownership by private capital interests is questionable, because it cannot be as stable or accountable as the state. For nuclear waste we need to take time. Yet our systems are not in tune with intergenerational needs nor the realities of sharing the planet. We are back to Greta Thunberg's cry of "How dare you!"

The power of oikos rises from how home and the household describe essential aspects of life. It reveals why dominant socioeconomic models fail to propose projects that will generate life. Oikos awareness is to insist economics, ecology, and ecumenical are related and always have been. Our charge is to ground economics in common human values and the reality of natural world, to change our systems and common agreements. Together they make the whole that is the household of the earth.

SCIENCE HAS SPIRIT

"Mommy, Daddy, how did I get here?" This classic question is good for us and Mother Earth.

The Roman Catholic geologist and paleontologist Fr. Pierre Teilhard de Chardin (1881–1955) gave some answers. You could see him as a sort of John the Baptist figure, to prefigure Jesus Green. I am astounded at how a priest who wrote major works before the Second World War still inspires and challenges.

Many have an aversion to the reconciliation of Christianity and science. Most of the objections should fall if you read Teilhard. He describes Homo sapiens as the outcome of life processes, how the myriad forms of life are not here by chance but the products of the quality of matter itself. He studied the ancestors of Homo sapiens, so he was ideally placed to describe how evolutionary force is present within us too. Teilhard is difficult to understand at times, but Julian Huxley wrote an introduction to Teilhard's

---

21. Canadian Nuclear Safety Commission, "Radioactive Waste," para. 3.
22. Waxman and Meadows, "Hearing for Radioactive Waste."

main work, *The Human Phenomenon*, which offers an easy summary. What you will miss, if you read no further, is the majesty of how we have come into being, like hill walking can let us look back and be amazed at where we have come from.[23]

Teilhard's description builds from first principles, from the nature of matter and simple cells. The same logic is applied to Homo sapiens. Teilhard is still heralded as a guide.[24]

What is most remarkable is how Teilhard looks into the future of humanity with some accuracy, based on his understandings of the past. He was among the first scientists to note a "drive to complexity" (*poussée de complexité*) in the universe, and he proposed this created a drive to consciousness.

There are steps in this history of life. Each provided the conditions for the next:

> *Cosmogenesis:* The big bang; the beginning of the universe
> *Geogenesis:* The beginning of the Earth, its crust, with tectonic plates and oceans
> *Biogenesis:* The beginning of life on Earth, which gives rise to the biosphere

Then, astoundingly,

> *Noogenesis:* The beginning of knowing we know; reflected consciousness or *mind* (from noos, meaning mind or reason)

Human consciousness has grown to the reality of global communications, the *noosphere*. The noosphere comes out of the biosphere, which relies on the geosphere. We easily recognize it in the World Wide Web and telecommunications, but Teilhard wrote in 1939. Science still had not described DNA. He was prophetic by knowing the characteristics and direction of the evolutionary processes rather than mechanisms themselves.

The technology of "intelligent" phones that give individuals the ability to load instant images and comments to this global space, called "cloud" sometimes, has fulfilled Teilhard's predictions. But Teilhard finds a further process at work, some of us find hard to believe, towards a superhumanity, the hyper-personal. This is an Omega Point (ultimate goal) of super life, with humanity reconciled to the whole of the biosphere, thanks to the noosphere, and enjoying hitherto unknown security and peace.

---

23. Teilhard de Chardin, *Phénomène humain*.
24. Danzin and Masurel, *Teilhard de Chardin*.

Put differently, h*umanity is incomplete; we have with science a new and powerful consciousness to integrate within the process of evolution, of which we have become a worker.*[25] Such an Omega Point lies far off. What I find more convincing is Teilhard's grasp of the process of creativity in the stream of life, arising from matter and energy itself.

He writes before the details of particle physics could be tested, but he is insistent that in the end all energies are psychic in nature. An amazing proposition, which matches the new physics of the Standard Model of how different forces can be found at the subatomic level. *"Atoms consist of particles . . . not made of any material stuff. . . . What we observe are dynamic patterns continually changing into one another—a continuous dance of energy."*[26]

Spiritual and material energy, the within and the without of things, are at play in evolution. How can they be dependent and independent at the same time? Yet we are familiar with the dual phenomenon of light as a particle and as a wave. Teilhard writes these are radial and tangential in nature.[27] He considers a slice of bread: "To think we must eat," but "what a variety of thoughts we get out of one slice of bread." Philosopher Alfred Whitehead (1861–1947) also emphasized the within and without of things. It gave theologian Walter Wink the basis for a best-selling analysis of Christian life.[28] Teilhard was unaware of Whitehead's work, but it is significant to find a scientist and a philosopher provided parallel observations through this critical period of the twentieth century. Einstein was not alone!

Teilhard found key themes to his majestic description: novelty comes out of restraint; the tree of life shows tendencies to diversity, complexity, and conscientization. As you read his works you may find yourself embedded in life, less afraid of stardust, more appreciative of lesser beings. He predicted knowledge with his reflection on the consciousness of plants: *"Is it not enough to see how certain plants trap insects, to be convinced that the vegetable branch, albeit from afar, is like the other two, subservient to the rise of consciousness?"*[29] How resonant with the insights of forester Peter Wohlleben and forest ecologist Suzanne Simard, who have described the interactions between trees.[30]

Teilhard makes you think, "Who am I? Who are we? What is it to be conscious?" His answers are contextual; we are materially dependent, but

---

25. My paraphrase of Danzin and Masurel, *Teilhard de Chardin*, 86.
26. Capra and Luisi, *Systems View of Life*, 77; emphasis added.
27. Teilhard, *Phenomenon of Man*, 64–65.
28. Wink, *Engaging the Powers*.
29. Teilhard, *Phenomenon of Man*, 153n1; emphasis added.
30. Wohlleben, *Hidden Life of Trees*; Simard, *Finding the Mother Tree*, 228–30.

material is more than we appreciate. To hold the within and the without together in awareness is like learning to swim for the first time. It opens awareness, freedom, and liveliness simultaneously, and Teilhard roots it in existence. If René Descartes is famous for the phrase "I think therefore I am," Teilhard may propose, "I think *because* I am."

The body makes awareness possible. Consciousness arises out of morphology: take Teilhard's illustration of a cone to explain how a small incremental change can affect a change of state. At the tip of a cone, in one increment, a section goes from a surface area to a point. Incremental changes in brains have continued in primates relative to the rest of the body, a pattern that repeats itself along all zoological stems. The movement to consciousness is not just human. Consciousness is the *"substance and heart of life in the process of evolution."*[31]

Teilhard asserts our human brilliance is to recognize relationships of all kinds. This prizes interdependence and interdisciplinary studies. It demands synthesis as well as analysis. We return to themes of ecumenism and economics! This is the new consciousness at the level of the noosphere.

André Danzin and Jacques Masurel suggest there is a new sense to Teilhard's use of the term *spirit-matter* and perhaps the extraordinary images of the early universe from the James Webb Space Telescope encourage it. The latest information theories suppose matter, information, and energy are the three main interchangeable components of the universe. The laws of physics governing the behavior of particles can be put as laws of information, tending to complexity, then giving rise to forms of "memory." This is the basis for negative entropy (negentropy) and order, rather than for collapse and disorder. It is not by chance that we are here. We can recognize Spirit in this significant order and share Teilhard's vision that "matter brings spirit."[32]

During my lifetime the understanding of evolution has moved on from a Darwinian emphasis on genes to include the dynamics of inner and outer realities, chiefly with epigenetics, which has a sort of example in the Old Testament when the crafty patriarch Jacob uses peeled bark to breed more striped sheep (Gen 30:37—31:16). External circumstances *can* affect internal processes, even switching on or off certain genes for multiple effects.

We are ready to acknowledge Teilhard's view that humanity is conditioned not only by genes but by the development of cultural and spiritual resources. He insisted that the human phenomenon is not found in the process of selection but in the continual emergence of spirit. He expected humanity to change and not simply to define progress by defending how

31. Teilhard, *Phenomenon of Man*, 178; emphasis added.
32. Teilhard, *Human Phenomenon*, 72.

we are today. We know we are changing and that the expanding means of communications has given humanity a sort of nervous system at the scale of the planet. Teilhard: "La Noosphere est une immense machine à penser"—a great thinking machine.[33]

Teilhard's writings disturbed both church and science. The Vatican of the thirties and forties resisted their implications for expressions of faith. Few scientists are comfortable with the degree of certainty with which he proposed the future. I agree. Studies of complexity show, even if we have a complete grasp of the elements of a situation, we can predict only a number of equally probable different outcomes. With prediction, there is also the danger of determinism. In his defense, Teilhard did not say which path we are taking to arrive at an Omega Point. He made anguished calls for people to engage in the evolutionary process, of which he sees them responsible: this is indeterminate. He emphasized education and the organization of science and technology to gather knowledge as our engagement in the evolutionary process.

Does Teilhard help us realize the threat to this hope comes not just from bad religion but from the commercialization of science and education, because science needs spirit?

YOU HAVE TO KNOW THE SYSTEM

This is not about an update to your computer! Systems thinking may seem an add-on to your daily life, but we are doing it already. If you travel internationally, you practice systems thinking to reduce the stress of your journey and present the right documents at the right moment. Outside the airport the systems can be overwhelming, with different languages, cultures, races, and very local phenomena; I remember my bewilderment with New York taxi culture in 1987 to arrive downtown without paying "an arm and a leg." I relied on a good friend who knew what to do.

What is true for humans comes from life itself.

A good example is the Gaia hypothesis, proposed by the atmospheric chemist James Lovelock and the microbiologist Lynne Margulis (1974). Five years earlier, Lovelock had presented his hypothesis of the Earth as a self-regulating system from his studies on volcanic eruptions and the atmosphere and how the Earth returned to a balanced state after major disruptions. The geogenesis that continues, such as volcanoes, and the biogenesis Teilhard described have interactive feedback mechanisms; the world is a living system of systems. The choice of Gaia as the name for this hypothesis

---

33. Teilhard, *Avenir de l'homme*, 220.

comes from a friendship between James Lovelock and the author William Golding (famous for his novel *Lord of the Flies*). Golding was researching the history of the Greek earth goddess Gaia at the time. He suggested it as the myth and reality seemed to match.[34]

This view of planet Earth is part of a paradigm shift in scientific perspectives: a change from seeing the world as a machine to understanding it as a network.[35] It belongs to a long history that shows changing understandings under the tension between understanding the parts and the whole. Modern scientific thought first arose as a science of qualities, of processes of transformation, with Leonardo Da Vinci, followed then by mechanistic science with Galileo Galilei and Francis Bacon.

The unprecedented problems of this century demand new levels of holistic understandings. Models of living systems can be explored, thanks to cybernetics,[36] complexity theory, and advances in computers.

In the case of Gaia, Lovelock developed a daisy-chain computer model of some fame: a coherent theory of living systems with a mathematical language is emerging.[37] These are the bases for deep ecology that include humanity as part of the environment: if we are just one particular strand in the web of life, then we need to know the outcomes of changing human activities for the whole. It is easy to appreciate this also has an inherent spirituality, given the sense of connection and belonging such ecological awareness brings.

Arne Naess made a distinction between shallow and deep ecology, where shallow ecology is human centered, technocratic, focused on pollution and resource depletion; and deep ecology is about asking deeper questions of value and belonging.[38] This radical sense also bridges into the existential basis of spirituality that I find so helpful in Teilhard's call for a re-expression of Christianity.

With this paradigm shift, all is reconsidered from a perspective of our relationships, including future generations and the rest of the living world. I immediately recognize the sixteen-year-old Greta Thunberg, lambasting the failure of the generations before her to make adequate changes. Deep ecology has taken root in her and the perspective of her peers—and far beyond. There are self-interested groups keen to dismiss deep ecology as too radical,

---

34. Capra and Luisi, *Systems View of Life*, 164.
35. Capra and Luisi, *Systems View of Life*, 4.
36. The science of communication and control in animals and machines.
37. Capra and Luisi, *Systems View of Life*, 12.
38. Devall and Sessions, *Deep Ecology*, xii.

when it is already present within millions of household discussions. We just feel powerless to make change.

With systems thinking we move attention from the parts to the whole, from objects to relationships, so multiple disciplines are required. Systems are nesting within systems, or ultimately, there are no parts at all. This brings a change in method from measuring to mapping. Mapping draws out patterns such as the networks, cycles, and boundaries that characterize living systems; we turn from quantities to qualities, from structures to processes, and from objective to epistemic science ("knowing" science, where the method of questioning is part of the theory it may propose).

Now if everything is connected to something, how is objective science still possible? Systems thinking allows shifts in perspectives as a complementary interplay, not as a replacement, so no perspective has dominance. There is room for "old" science, but not ultimately to turn to physics as "the answer to everything"; rather, this looks to life sciences.[39]

Perhaps better known than the Gaia hypothesis is the imagery of fractals. Systems thinking has led to the recognition of amazing patterns. This comes from chaos theory and contributions of mathematics to identify ordered patterns in chaotic systems, such as tornado behavior and how small events make big differences in their formation.[40] But most famous is the Mandelbrot set as the photomontage on the front page of *Scientific American*, August 1985, where seeing is "wondering." We are drawn into pattern repeating upon pattern, to the miniscule, and the recognition that mathematics contributes to our understanding of the living world. Fractal realities have been studied over the centuries with the geometry of plants and the repeating spirals in snail shells, snakes, and pine cones; mathematicians established the "golden ratio" and Fibonacci sequence. This is not by chance but by systems.

In response to the key question "What is life?," systems thinking focused on cells as a basic unit. This identified characteristics for all living things:

- Self-maintenance (or autopoiesis; self-regeneration from within)
- Non-localization (life is not localized but a global property of the cell)
- Emergent properties (isolate any of the cell's components and you will not find them alive; life is a combination)
- Interaction with the environment (a cell is an open system: although identifiable as a cell, it is also dependent on external realities)

---

39. Capra and Luisi, *Systems View of Life*, 15.
40. Bress and Gruber, *Butterfly Effect*.

The interaction of life with the environment is dynamic. It began with the creation of the atmosphere but continues with coevolving interactions. When this creativity is expressed by an organism, it is called cognition.[41]

So, something we take as a very human activity with cognition is an inherent trait in life itself. Humans have added recognition and reflection. The authentic frustration of Greta Thunberg before the global media of the United Nations was not just personal; it was phenomenal, like a sort of collective birth cry—explicitly political, out of the biology of the world. Greta was given the stage. How do we explain the inertia and impotence of institutions like the UN?

Humanity is in process itself, and a Green consciousness, even a shallow one, is emergent. A tipping point has not tipped. There is no consensus on the evidence for what percentage of the population makes change possible (10 or 25 percent), but tipping is real; like icebergs falling apart, all appears solid until . . .[42]

Given what we do is largely affected by our consciousness, it is important to know how human beings develop a new consciousness, and studies have shown there are predictable stages. This makes contexts like the USA, where the political choice is of two dominant parties, arguably a big problem. Multiple parties allow for the diversity of consciousness to be expressed in policies, and we see this in forms of proportional representation in Europe, where forty out of forty-three governments have some sort of proportional system.[43]

In 2021, Green parties shared power in coalition governments of six European nations.[44] Representation in public life of deeper Green perspectives is part of the dynamic of the paradigm shift we are living through. Knowledge of how people change helps to reduce disappointment or potential violence. It can guide us to take appropriate strategic choices in education, funding of government, and public policies.

Patterns of human development have been recognized for decades but not well taught. This may be as simple as the visible versus invisible, with a chicken-and-egg situation because the vocabulary is a little clumsy and remains unknown. A baby grows into an adult with predictable ordered stages, so puberty does not happen until a child has most of the abilities of an adult, but invisible things, like personality, beliefs, abstract thinking, and empathy, are part of this process too.

---

41. Capri and Luisi, *Systems View of Life*, 129–35.
42. Wagner, "Social Tipping Point"; Noonan, "25% Revolution."
43. Palese, "Which European Countries."
44. Nevett, "How Green Politics."

American psychologist and philosopher James Mark Baldwin (1861–1934) was among the first to identify distinct stages of human development. The Swiss psychologist Jean Piaget (1896–1980) refined Baldwin's work and gave the terms for cognition that are still with us: sensorimotor, preoperational, concrete operational, and formal operational. There are other sets of terms, but their similarities to Piaget's imply he established the basics. With each stage comes a set of values and a worldview, found across cultures. A large sample of an adult population will show subgroups of different stages of development. In 1995, American sociologist Paul Ray studied the role of values in American society and found they grouped around three large subcultures he called "traditionals" (25 percent), "moderns" (51 percent), and "cultural creatives" (24 percent), the latter emerging in the 1970s. We could name these as postmodernists. The subcultures corresponded with the stages of individual consciousness described by psychologists.[45] Popular culture knows other terms, like Gen X or Z, by age differences, but this ignores that across a generation of adults there is also a variety of values and worldviews that reflects these different stages.

Like Teilhard, McKintosh supposes human consciousness is creative, and that a new stage of consciousness has emerged as integral consciousness to surpass postmodernism. A well-known proponent is scientist Prof. Ervin László (1932–). He addressed the First Integral European Conference in 2014: "Integral consciousness is more able to relate to what is." The world is a global holistic reality, so better known by a global holistic awareness. "What you do is affected by consciousness," says László, "but what can one person do?" He encourages work on our inner realities.[46]

Ask yourself where your consciousness comes from, or simply acknowledge we do not ask this question often. László highlights near-death experiences, which challenge the dominant notion of a self-generated "turbine" consciousness that minimizes and separates us from the whole. Near-dead individuals' awareness was present despite their brain function being flat. Individual consciousness is also holotropic consciousness: we return to the butterfly effect and how each individual inevitably affects and belongs to the whole.[47] You will not be surprised that Teilhard de Chardin is named as an early integral thinker.

Another important contributor to integral consciousness is Ken Wilber.[48] He developed the concept of holons, or whole-parts, from the

---

45. McKintosh, *Integral Consciousness*, 66–67.
46. See László, "Integral Consciousness," 1:09; 1:30; 1:55; 2:20.
47. McKintosh, *Integral Consciousness*, 11, 82–83; Grof, *Holotropic Mind*, 17–21.
48. Wilber, *Sex, Ecology, Spirituality*.

Hungarian thinker Arthur Koestler with his novel *Ghost in the Machine* (1976). He argues holons make up the totality of the natural world, including humans. The living world is not a set of Russian dolls; it is more interactive and inseparable. Holons allow an exploration of the wisdom that we are more than the sum of our parts, as found in Aristotle's *Metaphysics* some 2370 years ago.

While it is easy in principle to recognize holons in some areas, it is hard to appreciate how they are a feature of everything and every dimension (almost). To help us, Wilber describes twenty tenets of holons that are fascinating to think about, and many of these features are already found in the systems thinking of Capra and Luisi. Tenet 9 is important for our environmental crisis: *Destroy any holon and you will destroy all holons above it and none below it.* We are foolish to undermine the integrity of the environment, as we are certainly a higher holon than insects, soil, plants, and ocean life-forms. Higher is more precarious. Wilber agrees with Teilhard's elevation of humanity, not so much as a leading edge of evolution but as vulnerable to the collapse of the holons beneath us. In other words, holons exist in a holarchy. Wilber phrased this memorably as: "Turtles all the way down and turtles all the way up."[49]

Integral consciousness is a growing movement and promises ways to cope with environmental challenges, not least by its appreciation of the value of other earlier worldviews (to be integrated, rather than surpassed). This allows for steps of change: a modernist consciousness will first embrace postmodern perspectives before embracing an integral consciousness; there is a holarchy in human consciousness too. The political art of social consensus for wisely based policy changes calls for integral consciousness. How else can systems be recognized and reconciled when in conflict? It seems the future of humanity awaits the arrival of a critical mass of integrally conscious politicians.

We do not have time for all of humanity to arrive as integral thinkers. Can a minority view lead us into a hopeful future—the 10- or 25-percent tipping point as a biological phenomenon? We are on alert for dangers and opportunities in cultural spheres as much as for food and shelter. Change has always come, despite sections of the population not fully understanding the reasons. COVID-19 policies relied on social acceptance, a level of civic trust, and they proved that trust was real.

Faith and spirituality are of great value in this integral view of life. When László said integral consciousness is "more able to relate to what is," he was very close to an important theological statement by Paul Tillich

---

49. See subtitle of Wilber, "Holons."

about faith. Tillich described faith as "courage to be" and defined God in an existential manner, as the ground of our being, or our ultimate concern.[50] His theological contemporary, Richard Niebuhr (1894–1962), understood faith as generated through our first experiences with those closest to us, good and bad, and in terms of the shared visions and values that hold human groups together.[51] All these levels operate simultaneously in the search for a center of value and power that is trusted and has meaning: the basis for faith.

For both Tillich and Niebuhr, faith is a universal human concern, not restricted to religious practice. We all have faith experience, waiting to be integrated into our worldview. So integral consciousness is a faithful consciousness, in its trust of wholeness, its emphasis on relationships, its knowledge of unity and meaning.

Rev. Prof. James Fowler (1940–2015) applied the work of Baldwin and Piaget to his study of faith development.[52] Faith has a story, a time line, a dynamic, close to the search for meaning that life's events disturb and stimulate. The church has confused belief with faith, reducing faith to doctrinal assertions whose meanings can have a sell-by date. Wilfred Cantwell-Smith (1916–2000), an expert in the comparative study of religions, was an important mentor for Fowler and gave him an understanding of many religious traditions and practices. He taught at McGill University and founded its Institute of Islamic Studies. I believe he would have enjoyed the Oikos Project! Just as with psychological development and cell biology, observation of religions revealed repeated patterns. Smith summarized faith with knowledgeable authority: *"Faith is what you set your heart on."* Vision is needed. Faith is a mode of knowing and expressed as a pledge of allegiance.[53]

Instead of the classic static question "Do you believe in God?," we now accept faith is more dynamic.[54] Personal faith development follows psychological patterns. It would be difficult to imagine otherwise. I find this helpful to remember as I prepare a worship service and question how it may help people at various stages of faith. Fowler claims earliest faith comes from a fusion of the child, the parents, and the cocreated values they share together, including inevitably the family history. Faith and the self are closely related. We can explore how this is lived out with a simplistic comparison: people who believe in many gods are able to live with a multiple sense of self, which

---

50. Tillich, *Courage to Be*.
51. Fowler, *Stages of Faith*, 4–5.
52. Fowler, *Stages of Faith*, 45–51.
53. Fowler, *Stages of Faith*, 11; emphasis added.
54. Tillich, *Dynamics of Faith*.

changes according to contexts, whereas monotheists (a classic Christian, Jew, or Muslim) have a dominant single self able to make sense of many situations. But belief in God is not Fowler's point; faith is needed by everyone and shapes their identity. Fowler describes the dilemma of the "henotheist" genius doctor, who is unable to integrate other aspects of himself and cannot stop being in role outside of the surgery. A henotheist would elevate one god while admitting others, and the BBC comedy series *Doc Martin* (2004–22) celebrates it!

Do I have faith in "an infinite source and center of value and power" that lets me integrate different identities, as a unified person, regardless of the roles and relationships I am in? Then I am a radical monotheist.

I have covered a lot, too much to avoid causing some confusion, but I hope this crash course through systems thinking brings on a sense of wonder, glory even, that life is as it is: that as we get closer to reality, our spirits soar.

In doing this research, I recognized my integral consciousness is in motion. Can we discern a "threshold" Green, somewhere between shallow and deep? Is there a tipping point of consciousness, where old values and ways of thinking are replaced by new ones, to the point that new meaning is found that is more hopeful? Would László say this is closer to the real?

In this change of consciousness, the drive remains the same, to make meaning and relate to the real. We are simply substituting understandings with new ones. These can grow through relationships of many kinds, come what may. They give us courage to be.

Deeper Green means you know the system.

THE BIRTH PANGS OF REAL MONEY

Like a visit to a planetarium that confronts us with the vastness of the universe, the multilevel and dynamic realities of the earth are overwhelming. It is easier to focus on one or two factors and explore how they belong to the bigger picture.

There is no better example than money. In Jesus' day it was a question of priorities because you cannot serve God and wealth (Matt 6:24). Now it is the definition of money itself that holds deeper Green potential. Things are already changing. You may have learned about online currencies and how to get rich quick, but in a Green book, we are interested in how money could serve a flourishing earth.

Sacred Economics: Money, Gift, and Society in the Age of Transition, by Charles Eisenstein, is an inspiring vision of how the phenomenon of money is key to "transition." The history of money shows it has changed, which means it can change again. Eisenstein tells a new story of money and the agreements to make with it.

The sequence of events matters. Public giving and receiving came before money. Money was created to make gift giving easier. More recently, nonmonetary activities, such as to bring water or offer childcare, have been taken into goods and services; monetization of everything makes everything scarce.

Eisenstein contrasts our not having enough time with how time is experienced in tribal cultures, or by each of us in our childhoods when days were long. We know how time can change, in an instant: the queue to pay groceries, or being trapped in an elevator, makes time long. Insomnia can make time painfully long. What of the time warp that happens in traumas like car crashes? Less dramatic: try a vacation on a barge to change the pace of life. Yet all these are taken to be exceptions, and time "flies" is the norm. *"We live in an abundant world made otherwise through our perceptions, our culture, and our deep invisible stories."*[55]

One *invisible story* includes the impact of Greek philosophy, and the distinction this made between the essence and the appearance of something. The real lay behind the appearance, as the essence, so the abstract became more important. Money changed from a sign of value to something that we take to be real in itself and, crucially, not subject to change.

The past and present economic reforms Eisenstein describes are therefore changes of consciousness, as well as practices. Money bridges between structures and values. It is time to question money and make it correspond to the knowledge of our place in the world. Eisenstein describes a system that rewards flow and not accumulation, creativity more than ownership. This corresponds with the planet, where every system is receiving, giving, and creating. It makes holonic sense!

Eisenstein considers the history of land ownership in stark contrast to this new vision.[56] Terrible stories of forced land ownership are easy to recall; the brutal seizing of native American tribal lands had its equivalent in Europe with the "freedom" of medieval serfs, by taking common land and "renting" it back. This "theft of the commons" paralleled what had happened with the creation of interest-earning money.

The Bible warns against usury, so it must be an old idea (Lev 25:37). It set off a cascade of consequences: if you borrowed to grow crops, you needed to produce more than before to pay off the loan. Debt creation

---

55. Eisenstein, *Sacred Economics*, 32; emphasis added.
56. Eisenstein, *Sacred Economics*, ch. 4.

sets pressure for growth of production and the sense of scarcity. We then measure an increase of goods and services to identify (exclusively) what is economic success (growth) with the term GDP.

It is helpful, vital even, to describe our money systems as *adolescent*: it was vital to grow, and now we are going through a coming-of-age ordeal. As adolescents we learned that we were unrealistic in behavior and understandings. So the challenge for our collective change is within the experience of each of us. We are happier as mature adults, or at least the angst of being an adolescent has gone. What might an adult money system look like?

Eisenstein follows a history of people who found the economic world unjust and questionable from a humanitarian perspective, because relationships matter. They insisted on the overlap of ecumenical and economic.

In 1879, the American political economist Henry George noted it was unreasonable for a landowner to profit entirely from the land's produce as land itself belongs to everyone and no one, so he proposed compensation: a land tax, to be paid by owners based on the land itself, not on the results of labor.[57] Nowadays, many other aspects of life that were given to humanity, like minerals, oil, water, and the electromagnetic spectrum, have been monetized in the same way as land. But "Georgian" compensation or land value tax (LVT) can still be levied on landowners. It is found in places as different as Denmark, Estonia, Russia, and Singapore.

A way to take things further is to change money itself. Eisenstein proposes a "currency of the commons." Government would issue money on a *science- and value-based consensus* on the limits to the use, or pollution, of natural resources. It would reclaim money to correspond with our collective consciousness. This currency would be redeemed by producers when they wanted to use or offset natural resources such as fishing cod or producing carbon dioxide. The more something was scarce or damaging, the higher the exchange rate. The currency itself acts as a tax on resources and pollution. Adjustments would be made year to year based on the secondary market of the item. The fossil fuel industry would have an incentive to find ways to keep oil and gas in the ground and reduce their costs.[58]

Another major change would be to offer credit at zero or negative interest, as a financial system parallel to the impact of a currency of the commons. At present, money is unnatural in resisting decay. It defies nature with the power of interest leading to an exponential growth of "value." Put this way it is easier to describe as an "adolescent" phenomenon. A stable

---

57. George, *Progress and Poverty*.
58. Eisenstein, *Sacred Economics*, ch. 11.

future demands a new money that is more natural. *"Decaying currency is no mere gimmick but an acknowledgement of reality."*[59]

Any of these changes would mean a reduction of taxation on sales and income with increased tax burden on material extraction and production. They are a practical means of integrating the "externalities" of production that are too conveniently ignored in classic economic descriptions. Those who benefit from the gift of resources for common human existence must also contribute to the larger community of life. Goods would rise in price while service costs would fall. Repair and recycling services, which value objects, would have new vitality.

Perhaps you would welcome a fall in taxation on your income but hesitate over changing interest rates on money. You are right that negative interest penalizes the storing of money and acts as a form of tax on savings. But such a vision is not new: in 1906 Silvio Gesell wrote of free money, where "currency decay" frees the role of money as a medium of exchange from money as a store of value.[60] It is also freed from control of the financially wealthy.

How unquestionably we accept that money makes money. Eisenstein claims: *"Money is so defining of our civilization that it would be naïve to hope for any authentic civilizational shift that did not involve a fundamental shift in money as well."*[61] To date I have found nothing as powerful to reflect upon, to make the transition of our times real and understandable. Eisenstein makes the case that we apply critical thinking to the nature of money itself as *part* of the Green solution. The mighty dollar can be like a golden calf: a dangerous idol if we take it as the foundation for life. Our values are changing. Money and its systems are more than ripe for reform.

We can welcome more than one form of currency; we know diversity provides resilience to species and ecosystems. Local and complementary currencies not only stimulate local economies, they also create community and neighborhood identities. This is not new, but often those running larger conventional financial systems viewed these as threats and took action against them.

There is a parallel history with the first electric car and the sad consequences of the dominance of petrol-fueled vehicles.[62] However, just like the electric car is rapidly gaining use, smaller-scale financial systems are blossoming. You can look up the differences between LETS, GETS,

---

59. Eisenstein, *Sacred Economics*, 224; emphasis added.
60. Gesell, *Natural Economic Order*.
61. Eisenstein, *Sacred Economics*, 214; emphasis added.
62. Matulka, "History of Electric Car."

Time-banks, fiat currencies, local currencies, local credit systems, coop banks and national banks versus commercial banks; the diversity reflects the range of people who have realized the link between the nature of money and the nature of community.

Eisenstein offers a third strand to *Sacred Economics*, after negative interest and money of the commons: the social dividend. This recovers one of the best aspects of pre-monetary civilizations, the gift economy. We still say the best things in life are free and set early retirement as one of the jackpots of a successful life. "Oh, to be free of work!" But what if we can experience more of this giftedness during our working life, because it is less defined by money, as well as money changing for the better?

I have witnessed the power of gift economy through my church's "Faith in Nature" program. We give away houseplants to increase short-term carbon fixing and talk about climate change. Recipients learn how to "grow it and pass it on," to manifest a benevolent pyramid system. To meet strangers at our table on the front lawn of the church and witness their pure joy at receiving a free plant has been one of the most uplifting experiences of my thirty years of ministry. It is multilevel. In an instant we break down many of the taboos of gift and scarcity that impoverish and oppress our present times. The conversations with this simple plant giving often lead into aspects of hope, friendship, purpose, and change. Gift giving reminds us about the true nature of wealth. Notice the gifts that come to you each day, as a new daily habit, and the effect of your noticing on your mood.

In *Sacred Economics* the future promises an obsolescence of "jobs" with a different experience of work: "Clearly, we face the means and face the necessity to grow less, to work less and to turn our energies toward other things. It is time to redeem the age-old promise of industry . . . and usher in an age of leisure."[63]

Leisure is not just to lie in the sun, as I write this book in a park, but any meaningful activities that will not necessarily produce salable products, with the freedom to choose your hours of "work." The disruption and suffering of the COVID pandemic has at least given proof of the flexibility of working conditions and proven an interest in shorter working weeks. One of the practical steps to a new era of leisure is the social wage (a 1920s concept from Major Douglas, the founder of the social credit movement).[64] This has already been practiced in a limited way with the USA stimulus checks sent to all households in 2008 and in Germany with the *Kurzarbeit*, or short week, to prevent unemployment.

63. Eisenstein, *Sacred Economics*, 274.
64. Eisenstein, *Sacred Economics*, 274–75.

In our original model of the household, that is planet Earth, I introduced oikos with three equidistant and undistorted circles; but the present situation is econo-centric. Like an egocentric attitude, this is usually a problem for individuals. The credo "I am my money" is known to be a mistake: there is Charles Dickens' Scrooge in "A Christmas Carol," and you will know someone for real. The scarcity around money can bring out the worst in cultures, as well as in people.

Stephen Sondheim wrote a musical about relationships in the music business, *Merrily We Roll Along* (1981). It is a tale of how the artistry of music, as well as friendships, could be suffocated by the lure of financial security. We know it is a problem, like playing with fire, and the regular headlines of fraud because of ambition prove it.

Even the films that are centered on money, to beat stock market crashes, rob the casino, or marry a millionaire, feature the artistry of success rather than a sustained love of money itself. The directors know the viewers would not be entertained or would even be made uncomfortable by brute materialism.

Perhaps you would not describe yourself as being money minded, but that does not stop the mass media from assuming so: there is an ideology of "economism" that we *are* our money. We *"the nation"* is OK when GDP goes up, so you can go to bed that night thinking the world is safe and going in the right direction. Yet the worse the environmental problems become, the clearer this is exposed as a propaganda, without a despot to remove. We are the propagandists by our love of the status quo. This is where a "happiness index" offers a critique and liberation.[65]

It is false to blame money for all our environmental problems and many others. But, Jesus had a particular interest in talking about money; I am on solid ground to finish a Deeper Green chapter with one of his most famous sayings.

"It is easier for a camel to pass through the eye of a needle than for someone who is rich [Gk: *plousion* = rich man] to enter the realm of God" (Matt 19:24; Mark 10:25; Luke 18:25).

This was a shocking contradiction to social assumptions and disturbed the status quo. People could point to the wealth of King Solomon (2 Chron 9:22), the restoration of Job (42:10), the promise of wealth from God (Mal 3:10). But Jesus meant a man's riches may not be a sign he is right with God. The absurdity is meant to reinforce the critique. It had huge ramifications because rich men were few in the context of Roman occupation. Most hearing Jesus must have laughed and maybe dismissed him as a dreamer. Yet it is

---

65. See ch. 7, p. 175.

easy to sense the same challenge from Jesus today, because the monetization of society and the role of finance are still male affairs. Jesus challenged male attitudes to money, then and now; we can have a good debate on whether a matriarchy would manage "transition" faster!

Our exercise to "follow the money" challenges any fatalism that nothing can change our path to self-destruction. It is to deny the facts. Money can change to serve a deeper purpose. Jesus answered the bewilderment that a man's riches would prevent his salvation—"Who then can be saved?"—by an appeal to God's nature to embrace all possibilities (Matt 19:25–26). But he was clear that wealth came with a moral liability. What will you do with this new knowledge of money? There is often vehement opposition to change, in the name of "freedom" and "free" markets.

The bigger picture overwhelms most of us. But we do not have to understand everything, not entirely, so much as trust because some *have* labored to grasp it. They describe how it relates to the great story of human civilizations and it makes our era exceptional—just as the paleontologists propose we live in the Anthropocene.

David Korton pioneered the presentation of this bigger picture with *The Great Turning*. The title says much. Korton had a varied career with development financing in Asia and then in nongovernmental organizations. His work through Harvard and the Ford Foundation with USAIDS gave him a knowledge of the mainstream, "the empire," and the clarity to present a global view of change.

The Buddhist eco-activist Joanna Macy understood this Great Turning from the perspective of working through the despair of a potential nuclear war; we are alive in the time of a third human revolution. The agricultural revolution took many hundreds of years, with huge impacts on human movement and population levels; the Industrial Revolution was faster, say three hundred years, with urbanization and even greater rise in human populations. It created an *industrial growth society* but on a finite planet; the emphasis on growth as a measure of success means civilization is out of balance. She describes this as a runaway system, and without balance it cannot be sustained.

In contrast, the vision for a *life-sustaining society* is already present. How can we move to this in a much shorter time frame? *Birth* is a better word than evolution for this paradigm shift; it allows for the pain of change and its rapidity. New money is part of a complex birth process.

Macy describes at least three levels of change, with an analysis in terms of flows of causality: one thing does lead to another. This allows for the change of scale and interrelatedness of things. It is a description I can understand without being an expert, which builds my resistance to fake, simpler solutions.

First, we can limit the damage of growth-based activities by holding actions such as tempering development with protective environmental laws, reductions to our consumption of fossil fuels, recycling, reusing, redesigning, rewilding, and replanting. But this alone is insufficient. Second, we need new structures, such as redefined holding of land, different measures of prosperity like new money, new forms of education that encourage awakening; this, too, is insufficient. Third is systems awareness: there are values and assumptions about reality that support new consciousness and ways of seeing the world. These honor ancient traditions of wisdom and native spirituality that speak strangely afresh. They allow for new knowledge about science and provoke a new sense of self.

I appreciate the interrelatedness and authenticity of this vision. A life-sustaining society has to be practical and full of meaning: any "revolution" must promise more life, not less. Individuals can contribute at different levels, and no one is burdened to do all. Macy says the Great Turning makes it an extraordinary time to be alive. The combination of these categories of causality means life can go on "with an exquisite reason to get up in the morning."[66]

Yet Macy was deliberately upbeat. Transition is painfully slow. For all the kindness and beauty I may offer in this book, the destruction and death caused by our "runaway system" of industrial growth still has us by the throat.

As a Christian, if I look to Jesus for guidance in this critical era, I experience an immediate disappointment: if we seek an easy, direct reading of Jesus, he is only "faint Green." Only when I researched deeper Green understandings did Jesus begin to resonate in ways that moved me in classic fashion. I recognized a sense of authority from his legacy. It gave me enough to dare to name Jesus "Green." Green as a title, because of what he means today, to those who wish to follow his way of life.

Now I must unpack my convictions, if you are to make a similar discovery: a Way that promises *home for all, not just for humans.*

---

66. Macy, "Great Turning," 5:17–20 (last three seconds of video).

Chapter 5

Jesus Christ Is Jesus Green

WHAT IS IN A NAME?

It takes a lot to change a name or to let someone take a new name. The Bible often equates a name with a state of being or action. If you know someone very well, a new name shakes up how we relate to them. Every time we trip on their new name or forget it, we sense something has shifted. Even subtly this raises doubts. Can I use their old name, even if they say they are fine with it? Do I know this person? Did I know them? Are we in the same relationship as before? What has changed, and am I somehow less connected? I am forced to reflect, and this includes a replay of the stories of how we met. It exposes there is to more know about the person, and anyway, how well do I know myself? Will I use their new name or not? It depends on the relationship, and time will tell.

When we welcomed a new canary to our household it sang the same day and interrupted our conversations. I felt a sudden pressure to find a name for this bundle of joy. I could not talk about a nameless bird and objectify a being who was already winning my heart: our red-and-orange canary is named Pyro, for fire.

Then Jesus Green can be named only as a response to the new relationship we make with his spirit, somehow carried by his story in the twenty-first century.

There are barriers: in my teenage years I went door-to-door as a youth club member to collect for the National Children's Home.[1] It has Christian origins, and often, as a way of saying, "No, thank you!," householders would say, "Sorry, I am not religious." I never argued back that the causes were

---

1. Renamed Action for Children.

humanitarian, but this is a key point for Jesus Green. It is time to get off the fence: his new name is not for the church.

This is Jesus who is "not religious" yet dealing with the ultimate; who can still speak to people as they skim the Bible the Gideons leave in hotel rooms. The name Jesus Green disturbs norms because of our common humanity and common dilemmas. We may be bombarded with commercial messages of perpetual indulgence to "Come and play! Enjoy yourself!," but nineties films like *The Truman Show* or *American Beauty* explored that route and how suburbia may be a dead end to our authenticity.[2] Sooner or later we face the fact we know we know: we will die; and we ask, "Why are we alive? What is our purpose?"

Jesus Green calls out against superficial living, as a foghorn, to avoid a wreck of civilizations on the undeniable realities of natural systems. But a book about Jesus can only sow seeds, and like his parable of a sower, it will depend on how the seeds fall (Matt 13:1–9). Some appreciation of what took place when Jesus was named Christ in the first place can prepare the ground.

Imagine you are in a catechism class and free to ask questions:
"How did Jesus Christ get his names?"
"Jesus was named by Mary and Joseph."
"Both names?"
"Well, no. He was named Christ when he was grown up."
"But how?"

The good priest will hesitate here. Will she avoid the easy answer? The temptation to take gospel truth as what happened is often too strong: let the teenagers sign on the dotted line at all costs.

I would rather talk about everyday titles. There is a nom de plume as a writer, actors take screen names, as a minister I am a reverend. There is a doctor of medicine or philosophy. Titles and roles go together but always through processes. People often pick out a peculiarity. In my school days we called one teacher Batman, another Skippy. Batman for his black gown, and Skippy as a wordplay on Hopton. This trait to characterize is present in religions, even those that deny any processes at all. If Jesus triggered some sort of recognition as a "Christ" character, he must have known it.

In an unusual conversation with his disciples, Jesus asks a question: "Who do people say that I am?" Given Jesus deflected attention from his healing miracles (Mark 1:40–44; Matt 8:1–4; Luke 5:12–15) and avoided the temptation to prove his identity in the wilderness (Luke 4:9–13), here he seems a little self-absorbed. But social anthropologists tell us first-century

---

2. Weir, *Truman Show*; Mendes, *American Beauty*.

personal identities came from the mesh of social relations to which one belonged. You could not define yourself.[3] It turns out this question was to prepare for the next one: "Who do you say I am?" . . . "The Christ of God," says Peter (Luke 9:18–22; Mark 8:27–30; Matt 16:13–20).

Although the Gospels sometimes portray Jesus as concerned to control his profile, he is wrapped up in power plays. He gave up his names with his crucifixion. Mark's Gospel hides and codes him until crucifixion and an empty tomb. Luke tells us God chose a young woman to name him, via the angel Gabriel. Matthew that he was born to it, as Son of David, Son of Abraham. John's Gospel places Jesus before anything had breath or form, and his crucifixion had the form of a coronation by Romans, who mocked Jesus as king of the Jews.

All the Gospels came through this defining death of a savior. Whatever names Jesus had as a charismatic healer and teacher had to make sense of his tragedy as well as the astounding claims of resurrection.

Christ is the Greek translation of the Jewish title Messiah, which means "anointed." There are many "christs" in the Old Testament. The best-known is a psalm of David:

> You anoint my head with oil;
> my cup overflows.
> Surely your goodness and love will follow me
> all the days of my life,
> and I will dwell in the house of the Lord forever. (Ps 23:5–6)

The Romans may not have known, but Hebrew kings were made only because God was willing to compromise! Once settled in the promised land, the One God wished people to be ruled through the collective decision-making of a group of judges. But people were jealous of the glory of their neighbors' kings. A "secular" pressure from elders of Israel pressured the prophet and judge Samuel for a change of mind, and they won out, even if God and Samuel knew it was a bad idea. It came with a formal warning (1 Sam 8:10–22).

By the time Jesus was born this warning had been born out through failures. Centuries of oppression with Greek and Roman rule, on top of national defeats from Assyrians and Babylonians, had broken open understandings. Kings were ambiguous at best. Yet there was hope for an anointed messenger, the Messiah, an agent of God's salvation. This hope drew on the book of Daniel (167–63 BCE):

---

3. Malina, *New Testament World*, 58–80.

> I saw one like a human being
> coming with the clouds of heaven.
> And he came to the Ancient One
> and was presented before him.
> To him was given dominion
> and glory and kingship,
> that all peoples, nations, and languages
> should serve him.
> His dominion is an everlasting dominion
> that shall not pass away,
> and his kingship is one
> that shall never be destroyed. (Dan 7:13–14)

How would everlasting dominion be possible? When Jesus taught "love your enemy" he was at odds with messianic expectations and popular hopes of an apocalypse to rid us of the Romans. He lines up better as Jesus the Winemaker than the Messiah, like his cousin John was the Baptist (John 2:1–11).

Events made a renaming possible, or even forced them. Jesus, crucified and risen from the dead, demanded a different sort of title than Jesus of Nazareth or Jesus, son of Joseph: Messiah was a name with the weight of religious tradition, in a good way, to admit the brutality and suffering he went through to become the hope and joy of his followers—the fact of failure, as well as goals of success. If Jesus refused titles, as an honorable man would deny being called good (Mark 10:17–18),[4] others insisted—just as I needed to name our canary Pyro.

The early church knew him as Jesus Christ; he was the Anointed One, and he certainly knew he was "in role" for the part when he went to his cousin, the Baptist (Mark 1:9–11; cf. Acts 4:27). The story goes that John publicly acknowledged Jesus as his successor and superior, so his baptism of Jesus was an anointing to prophecy, an accession, like Elisha received power from Elijah (2 Kgs 2:11–14). The prophetic Christ came before the royal one. Hebrew hope before Roman judgment.

The Gospels tell us Jesus was anointed twice: the Holy Spirit came on Jesus with baptism by John to power his prophetic mission. Nearer his death, a woman anoints Jesus with expensive ointment (Mark 14:3–9; Matt 26:6–13; Luke 7:36–50). Scholars question this story because it is laden with editorial meanings, like "preparation for his burial." It fits Jesus' antihero identity—his mixing with sinners, Romans, tax collectors, and fisher folk. Perhaps the expectations of a Messiah as a conquering hero make the case

---

4. Malina, *New Testament World*, 93.

that Jesus was first called Christ *before* heading to Jerusalem while skies were blue, and it stuck.[5]

The debate of how and when Jesus received the name Christ is about probabilities, not answers. It is the fact that Jesus was not born as Christ that gets my attention. Jesus Christ would have been a far less credible name than it now seems. There was a historical process. It means it is not so strange to call Jesus "Green."

A contemporary of Jesus could have helped us with this process. He, too, received a new name. The apostle Paul was a devout Jew named Saul, who held the coats while others stoned the first Christian martyr (Acts 7:58). Paul's silence on this (and much more) is frustrating, but he did leave an extraordinary appreciation of Jesus as a hymn quoted in his letter to Christians in Colossae:

> He is the image of the invisible God, the firstborn of all creation, for in him all things in heaven and on earth were created, things visible and invisible, whether thrones or dominions or rulers or powers—all things have been created through him and for him. He himself is before all things, and in him all things hold together. He is the head of the body, the church; he is the beginning, the firstborn from the dead, so that he might come to have first place in everything. For in him all the fullness of God was pleased to dwell, and through him God was pleased to reconcile to himself all things, whether on earth or in heaven, by making peace through the blood of his cross. (Col 1:15–20)

A hymn takes time to come into use and recognition, so this eulogy of Jesus arguably goes back to very early in the history of the Christian communities. It developed rapidly by borrowing other concepts. Do they come from the Jewish wisdom tradition of creation (Prov 8:22–31) or from widespread awareness of Greek stoic philosophy, with its metaphysic of God permeating all of nature?[6] Paul makes his points by drawing on the worship life of the church. Whether or not this cosmic reach was part of the original qualities found in Jesus Messiah, it was part of the very early church. Did it come even earlier, during Jesus' lifetime, if he was seen to be an embodiment of Divine Wisdom?[7]

By the time "John" wrote his Gospel, Jesus had become extraordinary (90–120 CE): "*In the beginning was the Word and the Word was with God and the Word was God*" (John 1:1; emphasis added). Jesus is the incarnate

---

5. Stuhlmacher, "Messianic Son of Man."
6. Balabanski, "Critiquing Anthropocentric Cosmology."
7. R. Fuller and Perkins, *Who Is This Christ*, 55–56.

and preexisting Logos, or Word of God. This seems a very Greek description if you know Plato and Aristotle, but Jewish writers Paul and the philosopher Philo also used Wisdom of God, Word of God, and Spirit of God. The Torah (law) was also understood as a manifestation of the Wisdom of God (Deut 4:5–6), and the Gospel of Matthew claims Jesus fulfilled it (Matt 5:17). Just as with Paul and his hymn for Colossians, John did not simply have a flash of genius. He drew on established truths.

These threads demonstrate the first and second centuries CE were far more fluid and open than we experience with religious ideas today. Syncretism, borrowing from several spiritual sources, was normal, and there was Christian attendance at Greek temples (1 Cor 8:4).[8]

A vital syncretism was at play in the early church. Can we welcome this again today to discover Jesus Green? It has been forty-five years since Fritjof Capra wrote *The Tao of Physics*, in which he noted the convergence of scientific and spiritual perspectives. Physicists and mystics come to the same conclusion about the underlying unity of things, from different starting points of outer realms and inner realities.

The incredible names given to the son of a Jewish carpenter tried to spell out the extraordinary demands felt by those who met him or heard his teachings. By faith, they have been recognized over the centuries, because in some ways the claims and experiences continued.

It is said that meeting God is a mountaintop experience. I think this is apt if it means being overwhelmed, awed, a little scared: I am not built for life on a mountain top. For God to be God, there is always a consequence to this meeting; I can borrow from Paul Tillich, and widen the language beyond theism, to say it is *an ultimate demand*.

The Bible has it right to bother about names. People get new names when things happen to them or they show new qualities.

Jesus can inherit a new name, as we bring twenty-first-century consciousness to his story.

THE DEEPER JESUS GIVES A SPIRITUAL WORKOUT

There is proof that Jesus is the Son of God: he was so wise, he never wrote a book! Seriously, we are told he took no steps to ensure his movement continued except by remembering his own death through a meal. What sort of strategy is that? Yet something happened to provoke his exaltation to a cosmic scale.

---

8. Barrett, *Gospel According to John*, 36.

What is it in the story of Jesus that so convicts and inspires? It is how it addresses universal questions. This is Jesus beyond religion.

Jesus confronted the universal questions of human existence in *his* own journey. What was his identity, his purpose, his origin? You do not have to agree with the answers in the Gospels, but you can notice these questions were being asked and answered in different ways. For three of the four Gospels, Jesus' struggle for answers was set in uninhabited lands, not far from the river Jordan.

There is an elemental dimension to the story, a stripping away of any social context except for being alive, dependent upon the earth, to see the open sky and stars—the experience of finding a time and place of ultimate demand. A classic spiritual summary of this existentialism is to label it an "I-Thou" experience.[9] The language of "Thou" signals a Reality beyond life as we know it. It has a quality that in the moment is overwhelming: "I am alive, to all this: if this is life, then what is it about?"

I challenge you to give yourself some time to marvel at the fact you exist, beyond words.

Matthew, Mark, and Luke are explicit that Jesus was driven by the Spirit of God into the wilderness for a time of fasting. This was a period of temptation by Satan with statements about Jesus' human-divine nature, identity, and freedom. It was also a preparation for the battle Jesus ultimately undertook against Satan and death on the cross. To "take up one's cross" as Jesus bid his followers can mean this sort of inner struggle.[10]

The "I-Thou" experience is also a quest in the humorous sci-fi novel *Hitchhiker's Guide to the Galaxy*, by Douglas Adams. The Magrathian civilization was dedicated to finding the answer to the question of "life, the universe, and everything." They built a massive computer, but the answer would take time, measured in generations. In the end the answer is forty-two, but no one has any details of how they framed the question, like finding a key but losing the address.

In real life, Adams resolved all sorts of speculations, some very clever ones, on how he chose forty-two. It was a joke. He wanted a smallish number, stared into his garden, and thought forty-two. Some computer![11]

There is no "answer" to existence, it just is. Being alive puts to us the question of "life, the universe, and everything." For Douglas it was his garden, not Jesus in the wilderness, and he knew he was kidding.

---

9. Buber, *I and Thou*.
10. See H. Williams, *True Wilderness*.
11. Adams, "Answer very simple."

In John's Gospel the existential crisis is resolved through a second birth of the spirit described to Nicodemus, leading to light and eternal life (John 3). It is powerful language, if you handle the symbolism. I recommend not gripping John's language too tightly.

A deeper Jesus can be found through the traditional church language of heaven, hell, forgiveness, and life everlasting, but the symbolism takes work, and most folk are too busy or wary others will assume they think it is literally true.

However, symbolism is still widespread these days, to address the question of the meaning of "life, universe, and everything"—just as nonrealistic art can be true, and often better than a photograph. Have you visited the Rothko exhibition in Tate Britain art gallery, London? Nine oversized canvases with blocks of somber grey and maroon colors have a likeness to Stonehenge. They reminded me of the arches of a cathedral, but the colors made me increasingly sad, and it was OK to own the sadness. It was real. Rothko gave me a place to take off some masks, with a genius to exploit how perception can connect with deep human experience. The work was created for an exclusive restaurant in a New York skyscraper, the Seagram Building, and is known as *The Seagram Murals*. The left-wing Mark Rothko was an unlikely choice of artist. He dined there, and it didn't go well. The paintings stayed in his studio, and we have a memorable quote: "Anybody who will eat that kind of food, for those kinds of prices, will never look at a painting of mine."[12]

The art of the Bible is authoritative when it paints existentially. Simply because science makes old language incredible does not mean old books were, or are, out of touch with truth. The language was never as literal as we may wish it to be.

Consider Jesus as a person of wilderness spirituality, with a cosmic reach. This stretches us upwards, outwards, and inwards, simultaneously. Think of monastic orders, with their intense worship life. There was transcendence in their music, but between praying the communal life had great demands.

Jesus moved on from the wilderness to teach in towns and villages, and ultimately to enter Jerusalem. He was not a city boy. He had an accent like Peter, as an outsider, culturally and politically. His teaching and travels turned upside down the notions of friends and family: his inner world. The deeper Jesus had wrestled with the "I-thou" experience, with ultimate questions. Now he engaged the "You-me-us" experience around Galilee.

---

12. Wolfe, "Mark Rothko's Seagram Murals," para. 7.

These vertical-horizontal and inner-outer dynamics to being human are inescapable.

I find this extremely helpful.

Being a very religious person is not popular. For many, perhaps most of us, friendships are what count, and big questions can be put to one side almost all the time. Earning a wage, the dramas of romance, friendships, entertainment, hobbies, playfulness, and escapism are enough. But this more horizontal realm is the scene for the majority of Jesus' teachings. He can come across as a great humanitarian, and his stories of the good Samaritan and the prodigal son are common knowledge. But then there is more.

The Jesus story is an invitation to the "I-thou" experience out of the horizontal routine, something you might expect with Jesus, as the son of God. The original contexts of his stories were electrifying, and the charge comes through human relationships.

Perhaps these words do not work for you. You are on a steep learning curve with this Jesus material per se, or if you are on the inside, church culture has not prepared you to think this way. Church can suffocate how to bring the deeper Jesus to life.

Consider a Bible story with a spicy context and some imagination. This is one of the most sensual stories of Jesus, which hinges on hospitality: a dimension that is key to the gospel of Jesus Green.[13]

All four Gospels describe how a woman comes to Jesus to anoint him (Matt 26:6–13; Mark 14:3–9; Luke 7:36–50; John 12:1–8). Luke's version tells us that Simon, a Pharisee, has invited Jesus to dine at his home. Pharisees were public figures, teachers of the law, who shared in the relentless game of how to maintain their reputation. To throw a feast was not to show you had money but to rise in honor by successful interactions with men who were more honorable than yourself. Given Jesus' notoriety, this evening could place Simon at the top of the town rankings: an occasion for as big a guest list as possible. Women were exempt from this scrambling for reputation; their realm was the home, but their behavior supported the status of their husbands.[14]

When Simon invited Jesus to his home, he took a calculated risk. Perhaps he trod carefully to show his invitation was more as a leader of his community than as a Jesus supporter, but it turned into a disaster.

Jesus got ready to eat, with some handwashing and probably a padded bench to lie on, when a "woman of the city, who was a sinner," arrived (Luke

---

13. Malina, *New Testament World*, 91–93; Wink, *Transforming Bible Study*, 138.
14. Myers, *Binding the Strong Man*, 198–99; Myers, *Sabbath Economics*, 24.

7:37), and anoints Jesus' feet with ointment. Her presence set off a series of questions to guests in the know:

"She prepared all this once she heard Jesus was at the house of the Pharisee. But how did she hear? How did she have the confidence to go to Simon's house? Was she a regular visitor?"

This sort of occasion was a semipublic affair, so her presence was tolerated. But to touch the guest, a holy man, the man of the moment, would create scandal, and she knew it. Crying, she wetted his feet with her tears and dried them with her hair, kissed his feet, and then anointed them with the expensive ointment. What could be a prelude for sexual intimacy is clearly spiritual devotion, in being so public and symbolic. Maybe Simon has seen this sort of adulation of Jesus in the streets, but this was in his home, on his reputation. Suddenly his invitation to Jesus did not seem such a good idea. Would Jesus stop her? *"Please, Jesus, send her away."* But that was in Simon's head.

The drama was what was *not* being said. Jesus reads Simon's, as if he had spoken out loud: "If this man were a prophet, he would have known who and what sort of woman this is who is touching him, for she is a sinner" (Luke 7:39). Finally Jesus speaks, "Simon, I have something to say to you." Was Simon relieved or in panic? Can *I* read Simon's mind? "OK, well, if Jesus wants to ignore her and talk to me, perhaps it is not so bad. Let's move on."

Jesus tells Simon a story of two debtors, both forgiven their debts by a creditor, but one has ten times the debt of the other. Had Jesus recognized a woman may sell her body to pay off debts? Jesus ends with a question, "Which of them will love him more?" Simon could have been insulted by such a simple question but he replies, "The one for whom he cancelled the greater debt."

This exchange is also something of a convention. It was not private, guests were listening in; it was more of a verbal joust, a competition, *and* Jesus' response to that unspoken drama: Simon offered no explanation to Jesus for the arrival of this woman to the meal table, who she was, even though Jesus would expect a comment (wouldn't you?). Jesus' question addressed Simon's ill ease, as well as his thoughts about Jesus' status as a prophet. Has Simon given an adequate answer? For the guests it seemed such a strange story and question. It was such a no-brainer, for who would be more grateful. What was Jesus driving at? Was Simon in great debt himself? You can sense that everyone held their breath for Jesus' next move, in a scene that was already going to be gossip for months.

"You have judged rightly," Jesus replies, and Simon groans with relief. Match null! It was just another bemusing preacher story. But Simon had jumped on a false hope, for up to then the "elephant in the room" of a female sex worker has been avoided.

Jesus says, "Do you see this woman?," and Simon's feast descends to disaster: "Are you kidding me?" But Jesus cuts deeper still. He compares Simon to her; Simon had not been hospitable, not washed Jesus' feet, nor dried, kissed, and anointed them, because he was still on the fence about whom he had invited to dinner; whereas this woman had grasped what Jesus offered. The gut punch comes when Jesus blasphemes in Simon's house because he says to the woman, "Your sins are forgiven!": a statement that stirs the guests, not just Simon. "Then those who were at table with him began to say among themselves, 'Who is this, who even forgives sins?'" (Luke 7:49). Jesus answers them combatively, again addressing the woman, which is to treat her as honorable among the guests: "Your faith has saved you; go in peace."

The settings for the message of the Gospels carry a depth that we can still reach; the woman anointing the feet of Jesus broke so many conventions. The story must have been told both to proclaim Jesus the Messiah and to celebrate the freedom he lived and died for. This was not because the story remained historically true (four Gospels give four versions) but because Christians had already discovered this experience for themselves.

The deeper Jesus means this discovery can travel in time: Do "life, the universe, and everything" promise you freedom and status, regardless of your past? Imagine if you know they do, come what may. That is the work of the wilderness.

Jesus of Nazareth appears to have exercised unforgettable powers to relate to God and to the world around him. The "I-thou" and the "You-me-us" experience of being alive come through the memories of his life and teachings, if we are willing to go deeper.

JESUS THE SYSTEMS THINKER

Jesus taught people to pray, "Lead us not into temptation," and for pastors like me this includes the temptation to imagine I can think like Jesus, like I just did! A focus on his teachings saves me: "He did not speak to them without a parable" (Mark 4:34a; Matt 13:34; Luke 13:18–19). To tell parables is to have a poker face before your hearers and insist they can turn over the cards themselves.

Some parables suggest Jesus had ecological insights, enough to make him faint Green.[15] But parables are found more widely in the Old Testament

---

15. See ch. 3, p. 73.

and ancient Middle East civilizations. A New Testament parable may not come from Jesus, and even when it does, it is affected by the Gospel writers and their communities. Matthew, Mark, and Luke drew different meanings from the same parable, through their settings. Yet Jesus' parables are really good and they stop; at least the parables in the first Christian writings are different, more allegorical. They lack the same openness. Perhaps Paul and other writers felt the storytelling parables were so much "the voice" of their Lord that it was unfitting to teach with them.[16]

Parables may let us get as close as we can to the historical figure of Jesus. I find the truth goes beyond this, and bear with me to discover why.

Jesus' choice to teach by parables appears to have been his distinctive strategy. I notice this as a Methodist, because the founder of Methodism, John Wesley (1703–91), was strategic in building a movement that became a church denomination: he made sure there was training for a band of lay preachers and set up classes of devotion, confession, and encouragement for attendees. These cell groups are understood to be a key reason Wesley made more long-term impact than his comparable leader George Whitefield.[17] Care was also taken that preachers read widely as part of their formation, including Wesley's set of forty-four sermons.[18] I have my copy, which includes instructions on hymn singing. A lot of method! How things had changed, because originally the word "methodist" was a put-down from Anglican clergy threatened by revivalist prayer and fasting in their midst. This "Holy Club" of Anglican clerics in Oxford practiced austere spiritual renewal.[19] The idea that spiritual growth could be achieved through a monastic type of personal devotion was ridiculed. Surely God moved more mysteriously than rigid routines of prayer. History proved skeptics wrong. The methods expressed a desire for a deeper spiritual life that John and Charles Wesley would discover. Their method derived from awareness and goals.

What does the strategic method of "speaking by parables" tell us? Can they offer us a lens on their author or the impact of their author at least? This is not Paul: no one is telling us what to think but *to* think.

Many parables are said to reveal the secrets of the realm of God, the central goal of Jesus' mission: "The time is fulfilled and the realm of God is at hand" (Mark 1:14). Yet parables make this goal enigmatic. They demand process. The story of "the sower" is called the parable of parables (Matt 13:3–17; Mark 4:13–25; Luke 8:11–17) and reflects the realities of sowing

---

16. Drury, *Parables in the Gospels*, 7, 37.
17. Rack, *Reasonable Enthusiast*, 191.
18. J. Wesley, *Forty-Four Sermons*; Cindy Wesley, "What Have the Sermons."
19. They were inspired by reading Law, *Serious Call*.

seeds: different growing conditions affect the outcomes and are compared to the response people make to "hearing the word." Mixed within the surrealist picture there is a touch of social rebellion: a hundredfold yield from the seed that fell in good soil would be joyful for any real farmers to offset the rent of greedy landlords.[20] It hints at how Jesus' identity is tied up with his mission for a new social order, a realm of God on earth. Mark's Gospel asks, "Has the seed fallen on good ground with you?"

The parable of the unforgiving debtor (Matt 18:23–35) takes us into the sphere of human relations without a mention of God, and yet it is all about human-divine relationships. Typically, there are several layers to the story: Are you hypocritical in how you want to be treated and how you treat others? Have you a true picture of your own troubles compared to other peoples'? What would the world look like if we all practiced mercy? Does indebtedness justify depriving people of their freedom and happiness? Are human relationships supposed to reflect the relationships we have with God? Do we block or encourage the realm of God?

The realm of God is like a mustard seed growing into a tree; a great banquet; a woman finding a pearl; a buried treasure uncovered. As a concept it came from Jewish prophecies, yearning for a time when God would be God again, for his chosen people. This is especially present in the genre known as apocalyptic, including the book of Daniel. But the parables of Jesus subvert easy assumptions, even ones the Gospel editors may have had, longing for the return of Christ in power and glory. The realm of God could be in the present as well as in the future (Luke 17:20–21) or known in healing and exorcisms (Luke 11:20). It was scandalously within reach: "Blessed are the poor in spirit for theirs *is* the kingdom of heaven" (Matt 5:3; emphasis added).

The parables that are out of line with the Gospel writers' presentations of Jesus suggest a respect for the kernel of his teachings. They had priority, maybe because they were widely known.

This makes the place of nature in Jesus' sayings of the realm of God significant. Even the pearl of great price is a product of nature. The processes of nature are taken to explore human reactions and relationships: seeds sprouting, fighting weeds, fig trees being fruitful or not. There are processes, setbacks, even fragility and waste; the imminence of nature and the present tense of God's realm go hand in hand: "Look at birds and flowers: do not be worried, the realm *is* here."

But this is not the strongest point for us, rather, it is the "forensic" access it may offer to the character of Jesus himself:

---

20. Jeremias, *Parables of Jesus*, 150.

> For the kingdom of heaven is like a landowner who went out early in the morning to hire laborers for his vineyard. After agreeing with the laborers for a denarius for the day, he sent them into his vineyard. When he went out about nine o'clock, he saw others standing idle in the marketplace, and he said to them, "You also go into the vineyard, and I will pay you whatever is right." So they went. When he went out again about noon and about three o'clock, he did the same. And about five o'clock he went out and found others standing around, and he said to them, "Why are you standing here idle all day?" They said to him, "Because no one has hired us." He said to them, "You also go into the vineyard." When evening came, the owner of the vineyard said to his manager, "Call the laborers and give them their pay, beginning with the last and then going to the first." When those hired about five o'clock came, each of them received a denarius. Now when the first came, they thought they would receive more; but each of them also received a denarius. And when they received it, they grumbled against the landowner, saying, "These last worked only one hour, and you have made them equal to us who have borne the burden of the day and the scorching heat." But he replied to one of them, "Friend, I am doing you no wrong; did you not agree with me for a denarius? Take what belongs to you and go; I choose to give to this last the same as I give to you. Am I not allowed to do what I choose with what belongs to me? Or are you envious because I am generous?" So, the last will be first, and the first will be last. (Matt 20:1–16)

Whom do you identify most with in the story? Who provokes you most and how? Can you put the story into your own words? Reread this after you tell your own version, and see what you changed or missed out.

The parable is not realistic, yet it does address reality. When you see people standing doing nothing on *your* street corner, what do you see? Do you support a social wage? With experience of the COVID-19 crisis do you think a little differently about unemployment?

Why were the people who are in the marketplace at five o'clock not hired already? There are many possible explanations, not just ones that suit our point of view: they are drunks, or the manager preferred to hire people he knew, or they were there all day anyway, waiting.

If you were in the story, as the first hired and last paid, would you have anything to say? Would you answer back the landowner who says, rather bully-like, not to complain about his generosity? How much do you think made a daily wage? The answer is key. If this wage was just enough to live by, how does that affect the story? To pay latecomers less, the expectations you might have if you worked all day, would be to leave workers and their families hungry.

This exercise takes us into an appreciation of the unsettling change that Jesus was bringing, for the better. It is a gritty story. There is a subversive critique of the status quo, through the power of generosity, which exposes a systematic injustice only some would recognize.

What does it tell us about the sort of person who would choose to teach in this manner and who would construct such as story? We could study many parables and find the same provocative and open-endedness emerges. A parable is a sort of verbal sticky ball. You might recognize a sticky ball and pick it up knowing it's going to be tricky to get rid of it, or it might take you by surprise. All the same, it is a process to shake it off: a parable is mental, not physical. It is not just any old story; it has been crafted. If we truly pick it up, it disturbs, every time.

Jesus repeatedly exposed the way each person internalized their participation in a system of domination and exploitation that can be resisted in the name of God.[21] Jesus' parables were part of a prophetic stream that welcomed a new order while the old order crumbled.[22] Parables are designed to transform rather than inform, as parents hope their quarrels with teens over curfews will become internalized awareness to the needs of others.[23] The impact of the famous parable of the good Samaritan is "to enable the listener to experience post-conventional compassion. To be told that one *should* be compassionate is not transforming, to *experience* compassion is."[24]

Just like the effects of parables themselves, studies on them continue to feel fresh and authentic. They point back to Jesus as a person of great awareness: an awareness-raiser has to have awareness. Stories crafted to provoke; to throw off oppression, to encourage hope and trust, to appreciate the now and the given indicate an exceptional consciousness. Jesus was a thinker who brought systems to light and equipped people to do the same.

One of the benefits of parables is an escape from linear thinking. Parables encourage us to question whether human progress will happen just by easier access to information. They expose a contemporary arrogance to suggest ancient people could not have been more aware than we are. Nobel prize-winner and physicist Leon Lederman would agree. He wrote an entertaining description of particle physics that appreciated the Greek

21. Wink, *Engaging the Powers*, 51–59, 82.
22. O'Murchu, *Quantum Theology*, 114–15.
23. Kegan, "What 'Form' Transforms," 64–65.
24. Spear, "Transformation of Enculturated Consciousness," 371; emphasis original.

philosopher Democritus (c. 460–370 BCE).[25] Democritus deduced the existence of atoms, millennia before they could be recorded and measured in a particle accelerator. Lederman found Democritus inspiring for his work to understand the forces and particles of nature with the discovery of the muon neutrino.

Scientific method is still working to catch up with the explorations of human consciousness, and systems thinking precedes systems theory. Jesus was not alone; most of the surviving fragments from Democritus address the questions of ethics rather than the makeup of the universe.[26] Why would they be separate anyway?

I invite us to take Jesus as a systems thinker, in his terms, just as we can be systems thinkers only in ours. In fact, the terms Jesus worked with come to us in our "hour of need."

Consider what consciousness is present in a teacher who points to the smallest thing, a mustard seed, to explain the greatest thing, the realm of God. His ethical rigor, which was so disturbing and inspirational, was based on a profound sense of relatedness. This is a prescientific worldview limited by the nature of the Gospels themselves, but we still find an emphasis that resonates with contemporary Green consciousness. Jesus' particular focus and approach, to work through domestic life with a fatherly God whose forgiveness goes beyond common sense, make it easy to assume Jesus would insist on the oikos model and more.

MY FATHER'S HOUSE . . . A DEN OF THIEVES

Of all the events in the life of Jesus, Jesus' actions against the temple of Jerusalem are key. They bring the deeper Jesus to our awareness because he knew they would have terrible consequences. His courage, or folly, came from a depth of faith. From his analysis came action, based on issues that are very familiar.

The 2019 fire of Notre Dame Cathedral in Paris demonstrated how threats to national monuments have a huge reach. But the temple was more important to Jewish life than even Notre Dame is for Paris and France. It entwined Israel's economy with its religious faith and was associated with the pinnacle of Jewish nationhood.[27]

The temple had taken over the looser spiritual practices of the bloody times of Joshua of the occupation of promised land. Settlement in a new

25. Lederman with Teresi, *God Particle*, 25–63.
26. Duignan, "Democritus."
27. Jeremias, *Jerusalem*, 56–57.

land meant there could be a place for the holy symbols of Jewish faith, especially the tablets of commandments given to Moses on Mount Sinai (Exod 19). The decisions to build a temple, by whom and when, relied upon God's blessing. Handsome King David was blemished by lust and power, but his son Solomon, who chose wisdom, was entrusted with the instructions for the building (2 Sam 7). Details of its dimensions recall the instructions for an ark given to Noah, to house animals and Noah's family. God designated a building for specific functions to facilitate good "I-Thou" relations, and thereby divine order in "You-me-us" relationships. A glorious building, worthy of Yahweh, replaced a "tent of meeting" in the wilderness. This house was to restore the entire creation to good order and echoed stories of God walking with humanity in the garden of Eden. (Ps 24; Gen 3; the starry promises to Abram, Gen 12).

The Hebrew people did not find life so easy in the promised land. Succession to King Solomon was resolved only by the creation of two kingdoms: the Northern Kingdom of Israel with the ancient sanctuary of Shiloh (taken by the Assyrians in 701 BCE) and the Southern Kingdom of Judah, where Jerusalem was located, then left alone and defiant.

The discovery of an ancient scroll in the Jerusalem temple (622 BCE), supposedly written by Moses, led to radical religious reforms endorsing the temple: regional temples and shrines were closed to bring spiritual and political power into one place. A version of this is known as the book of Deuteronomy.

A further catastrophe struck in 587 BCE when the latest superpower, the Babylonians, brought national Jewish freedom to an end. Solomon's Temple was destroyed and rebuilt only as part of the slow return from exile.

The Second Temple was still a magnificent building:

> Now the outward face of the temple in its front wanted nothing that was likely to surprise either men's minds, or their eyes. For it was covered all over with plates of gold, of great weight: and at the first rising of the sun reflected back a very fiery splendor and made those who forced themselves to look upon it, to turn their eyes away: just as they would have done at the sun's own rays. But this temple appeared to strangers, when they were coming to it at a distance, like a mountain covered with snow. For as to those parts of it that were not gilt, they were exceeding white. On its top it had spikes; with sharp points; to prevent any pollution of it by birds sitting upon it. Of its stones some of them were forty-five cubits in length, five in height, and six in breadth. Before this temple stood the altar, fifteen cubits high; and equal both in length and breadth: each of which dimensions

was fifty cubits. The figure it was built in was a square: and it had corners like horns; and the passage up to it was by an insensible acclivity. It was formed without any iron tool: nor did any such iron tool so much as touch it at any time. There was also a wall of partition, about a cubit in height, made of fine stones, and so as to be grateful to the sight. This encompassed the holy house, and the altar; and kept the people that were on the outside off from the priests. Moreover, those that had the gonorrhea, and the leprosy, were excluded out of the city entirely. Women also, when their courses were upon them, were shut out of the temple. Nor, when they were free from that impurity, were they allowed to go beyond the limit before mentioned. Men also that were not thoroughly pure were prohibited to come into the inner [court of the] temple. Nay the priests themselves that were not pure, were prohibited to come into it also.[28]

Jesus was alive in the Second Temple period, which ended in 70 CE when the Romans destroyed the temple to defeat the Jewish rebellion.[29]

You might have a picture of Jesus as a generous and tolerant man, but the Gospels give us a moment of hot anger when he addresses the staff of the temple as "white-washed tombs" (Matt 23:27–28). At least Matthew thought Jesus would have said this. Did Jesus believe Daniel's prophecy that abomination was found in the temple (Dan 9:27)? His criticisms match the report of a nine hundred–fold reduction in the costs of doves during the Jewish War (66–70 CE).[30] That was some markup!

Even for a charismatic itinerant prophet, the temple took central place in Jewish identity. Jesus' primary message to repent and believe good news of a new realm of God set him against the status quo, so there was no better way to focus attention than action against temple corruption. He declared it was beyond repair (John 2:19; Matt 26:61; Mark 14:58). The Jewish-Roman power relationship had no capacity to respect "my Father's house." The status quo bristled with complexity from the simple truth: Rome was robbing Jews, and the elite Jews passed it on.

If Jesus' act of temple outrage against money changers was the deciding one in forcing the hand of temple authorities to arrest him, then we may have the moment in Jesus' life that he loses stability. Some commentators suggest Jesus was mad to follow such a futile path; even the Gospels tell us he seemed possessed (Mark 3:21–22; John 10:19–20). It is a rich debate.[31]

28. *Wars of the Jews*, 5.5.6, in Josephus, *Complete Works*, 848–49.
29. Anderson, *Living World*, 519–20.
30. Myers, *Binding the Strong Man*, 301.
31. Meggitt, "Madness of King Jesus," 379.

I side with the sanity of Jesus, to justify his actions. He knew what he was doing; it was consistent with his message as systems thinker who worked in parables. Jesus reflected deeply on ordinary life and challenged "normal" values. Parables disturbed the conventional thinking that was normal to New Testament times. Jesus knew he had to confront the principles and practices at the heart of the nation in multiple ways. His healings caused debates over who could heal, where, when, and how. He did not just turn over tables; he challenged the interpretation of the law, that it was what you did, your motives, that made you impure, rather than what or how you ate (Matt 15:10–20; Mark 7:14–23). The conventions of purity were closely related to the temple sacrificial system; to cheat the majority, the poor, at the moment they sought forgiveness undermined any honor for the status quo.

The house of God had standards that were inescapable. Jesus knew the demands of the Torah included how women and children should be treated, and how forgiveness was offered for all, not to a privileged few. The stories of Jesus suggest he believed God's forgiveness took priority over questionable religious practice. This could be at home, rather than the temple precincts. The conflict is clear, but did Jesus expect to die *or to be saved*, as if he were Elijah in a chariot of fire (2 Kgs 2)? We do not need to know, because he has given a message: his Father's house was meant to support what was of ultimate importance, not to exploit it.

It is fun and easy to speculate on a time-traveling Jesus, but the Gospels portray him being grittily concerned about everyday life, with surprising generosity sown into the fabric of how we treat one another.

Even today, this makes waves: Are we not thieves of the house of God that is this planet Earth? Rhetorical? Jesus' contemporaries would have found replacing the temple system just as problematic as our Greening the economy. Jesus understood human motivations, what it meant to be trapped in systems and still face material necessities. When the temple was destroyed his followers were able to follow him through its debris because they had a greater system than Rome or temple: Jesus and the realm of God, where God is at home.

This Jesus offers resilience for our times. Stories of God's realm address the hard realities of the daily wage (Matt 20:1–16) and the extraordinary power of natural processes (Matt 13:31–32); that we judge others less and live with difference (Matt 13:24–30).

It is easy to place the parables of Jesus into the oikos model of deeper Green.[32] He had owllike powers of observation, which crossed categories:

---

32. See ch. 4, p. 87.

someone today who would expose the contradictions of the status quo, and point to safer, truthful change.

THE EMERGENT JESUS GREEN

Few of us have asked how Jesus became called Christ. We take it for granted, but the process was one of emergence.

The Gospel writers and the early church claim that Jesus' true identity is known and shown in the circle of his followers. Peter's declaration to Jesus, "You are the Messiah, the Son of the living God" (Matt 16:16), could have historical truth because the title "Christ" was given to Jesus very early, almost as a proper name.[33] I would like to ask Peter *why*.

I love wine, especially the freedom to choose a bottle, like books in a library. Some choose the first front label they see that pleases, but anyone with experience realizes it is a con and reads the back label, with the origins, vintage (year), grape varieties, color, nose (fragrance), and flavors. The words are specific and relative. Vanilla is something we could agree on, but what is "round in the mouth"? The answer demands many bottles of pleasure. Other bottles offer no description at all. The name alone is supposed to impress. Wines are complex realities with more than can be described. Hence the pleasure of choosing and drinking them: many things come together, something new emerges on your palate and perhaps in a name.

Emergence is a key phenomenon for this book. It means "the arising of novel properties in an ensemble, novel in the sense they are not present in the constituent parts."[34] Emergence has become a buzzword, but it dates at least from the mid-1800s with British "emergentism."[35] Now it has been recognized as a fundamental phenomenon, found in very separate phenomena, from behavior in cells to social theories; in life itself to online controversies. Ken Wilber's elaboration of holonic theory is a specific way to understand how emergence is found everywhere.[36]

Think of the extraordinary people of our times, with such charisma they could hold an audience spellbound. When they die the impact of their life changes. Elvis was the living king of rock and roll and crowned by his early death. Jesus was not named as "Messiah" but "Christ," its Greek equivalent. Did his title emerge as the message went to Greek-speaking Jews and Gentiles? It was certainly a vacancy people wished to fill. If Jesus

33. Dunn, *Theology of Paul*, 197.
34. Capra and Luigi, *Systems View of Life*, 133.
35. Mill, *System of Logic*.
36. See ch. 4, p. 98–99.

was first named *Mashiah* (Hebrew: Messiah) with expectations of being a divine hero, he was a disappointment. He entered Jerusalem on a donkey, not a horse, taught "Blessed are the meek," and advised "you should take the worst seat at a banquet" (Luke 14:8–10). Perhaps "Christ" was an easier title, because in translation it lost the baggage of Hebrew expectations, or even disapproval from Jesus himself. Given Elvis resisted being called king, how much more would Jesus have frowned on being called Messiah? Perhaps the title came from his own followers, like the Gospels claim.

We can never know the exact process of the emergence of Jesus named Christ, but we can make use of the earliest writings. They show a dynamic between the written word and human activity, which repeats itself with this book.

The seven letters of Paul, and two others in similar style, use Jesus Christ without explanations. These are commissions and commands for groups of Christians, mostly beyond Israel, at least a decade or two after Jesus' death, and they include arguments, dressings down, denials of rumors, and defense of his credentials. That Christ is "taken for granted" does not mean "couldn't care less."[37] Nowhere does Paul justify calling Jesus "Christ." Neither does he give much detail of the life of Jesus which seems very strange. We have hints in how Paul portrays Jesus and in what he finds important for the church. One explanation is that stories, such as Peter's declaration, were already taught orally and part of a shared expectation Jesus would return in peoples' lifetimes. These letters fail to let Jesus Christ emerge because they had other goals.

Paul uses other titles for Jesus: *the* Christ, to make the point of his messianic identity (Rom 9:5); the Spirit of Christ (Rom 8:9); the Lord, or our Lord (Rom 1:7; 1 Cor 1:2–9); the Son of God (Rom 1:4); the last Adam in the sense of his whole human life (1 Cor 15:45), and the preexistent Wisdom of God (Prov 8:22–31, in 1 Cor 8:6; 1 Col 1:15–20).

I chose wine to illustrate emergence given Jesus' famous first sign of water turned to wine at a marriage in Cana, and his remark that new wine is for new wineskins (John 2:1–12; Mark 2:22). I hope Jesus would approve. We can never know exactly when people claimed Jesus was divine nor precisely what they meant, but after the crucifixion and burial of Jesus, he was experienced differently: as alive in a way that determined his titles.

People reassessed Jesus and shaped the content of "Christ." A new Jesus emerged from claims of his resurrection as he went beyond the frontiers of human existence: the Jesus of history became the Christ of faith.

---

37. Dunn, *Theology of Paul*, 185.

What language could be adequate to convey the claims about Jesus? Theologian and philosopher Paul Tillich took a functional view of the names for Jesus. Religious titles, like Christ, are symbols of faith, and only such symbols can express the "ultimate," a.k.a. God. *But how do symbols function?* The answers can help us to understand some of the processes behind the names of Jesus, beyond religion. Tillich described symbols by six characteristics and distinguished them from signs.[38]

Symbols share their first characteristic with signs: both point beyond themselves. But a symbol participates in that to which it points: the flag participates in the power and dignity of the nation for which it stands. It cannot be replaced except after a historic catastrophe that changes the reality of the nation which it symbolizes. An attack on the flag is felt as an attack on the majesty of the group in which it is acknowledged. Such an attack is considered blasphemy.

A symbol opens new levels of reality. The art of a picture and a poem can be a symbol that reveals elements of reality that cannot be approached scientifically. In the creative work of art, we encounter reality in a dimension that is otherwise closed for us. I think of Auguste Rodin (1840–1917), who titled his two-figured sculpture *The Kiss*. It manifests the interiority of the moment.

The symbol's fourth characteristic is how it reflects the human: it unlocks dimensions of our soul. A great play brings up hidden depths of our own being. We have dimensions within us of which we cannot become aware except through symbols. A parallel is found in the melodies and rhythms of music. What is your favorite tune, and why?

Symbols cannot be produced intentionally. They grow out of the individual or collective unconscious. Symbols that have an especially social function, such as political and religious symbols, are accepted by the collective unconscious of the group they belong to. (This makes symbols emergent phenomena: I found a seventh feature.)

A final characteristic of symbols follows: like living beings, *they grow and die*. The symbol of the "king" grew in a special period of history, and it died in most parts of the world in our period. Symbols do not grow because people are longing for them, and they do not die because of scientific or practical criticism. They die because they can no longer produce a response in the group where they originally found expression.

A comment from Rabbi Aaron Flanzraich is easier to remember than a list of features: "Signs tell us to do things, symbols help us feel."[39]

---

38. Tillich, *Dynamics of Faith*, 47–49.
39. Guevara-Mann and Flanzraich, "In the Beginning," 3:50.

How does Jesus Green make you feel?

First, he must grow!

Perhaps you take Jesus Christ as real, not a symbol. His title is revealed in the salvation history of the Jewish people and bound up with God's very existence. But Tillich's focus on function is not exclusive. He does not confront conventional beliefs. Rather he adds new understandings to our use of religious words, to make sense of the stories of Jesus in contemporary terms. God is our "ultimate concern," and faith in God is our "courage to be." I find Jesus' story tells me about our "ultimate concern," and that it carries the spirit of Jesus.

Jesus Green emerged in my consciousness when I brought deeper Green knowledge to my awareness of Jesus' story. Jesus Green is not my Green director; he is the symbol for our Green humanity, he has a place in our paradigm shift, the Great Turning, the Greening of human civilizations. The impact of his teaching, dying, and rising can speak to our needs for repentance and new behavior, to be reconciled with nature.

In this logic, the title Jesus Green can work only by our consent. Jesus and Green must trigger new reactions by their combination as symbols. One analogy can be of building a bridge and the moment the final midsection connects one side to the other: both sides of the bridge are affected by its completion.

Ecologists have named keystone species. Jesus Green may be a keystone symbol. We live by him to avoid self-destruction. But how? The remaining chapters will keep the tension between history, science, and faith and demand your reengagement with a story: a consideration of a singular tree where everything should have been dead and buried. Where Jesus *Green* says, "Father, forgive them; for they do not know what they are doing" (Luke 23:34).

Chapter 6

A Passion for Green

IMAGINE

The Beatles pop group gave me sad and funny memories from my student days in Liverpool. I was thrilled to make new friendships in the intensity of my first term. On Monday, December 8, 1980, I went to say goodbye to a classmate. When he opened his door, I saw two friends sitting in a vacuum of words, "John Lennon has been shot!" It was the sort of shocked sadness that came decades later with the death of Lady Diana. Where were you?

Both people made us smile, so another memory balances the story. The nearest Methodist church to my first-year residence was a short walk along Penny Lane. It was also the students' church, so "Penny Lane" is in my heart! The song's title comes from the bus terminus at the end of the lane where Paul McCartney changed buses to get to John's house. On Sundays I would rush up the lane but rarely got to church before the first hymn had started. As we sat to pray, I would get a nauseous but familiar smell. There were lots of leaves in Penny Lane that autumn, hiding dogs' doings, and we quickly named it "Shit Alley."

My first months away from home, aged eighteen, were exciting. "Who am I? What do I really think?" People I met made a big impression, and most of this was around passion: the passion of teaching, or desire, or friendship. It is not so different than the dramas of Jesus' intimates who looked back and tried to make sense of things. Imagine. I find passion in the crisis of planetary destruction and the eyes of those who see change for the better. There is joy in a love for the earth, to put a sparkle in the eye, not just tears, and I tell you these humdrum stories to connect ordinary and extraordinary vision. We can imagine a new earth with passion for Green, and Jesus can dance in our lives, and beyond our dying. Life and death, jokes and creativity, with the sense of eternity.

Perhaps I have already made you remember one Beatles song. I apologize to most of you who love it, *a lot*, but John Lennon's song "Imagine" makes me uncomfortable. Lennon painted a memorable vision of peace. I want more. Are we changed by singing it together, or are we tripping out? This was a valid criticism of the sixties and hippie culture. Yet some did act for change, as we have seen with the progress of environmental and human rights movements. It was a time of violence by states and terrorists when imagination breathed hope. Sung out of context today, I long for a second version, to celebrate *how* this vision can be made real.

Imagine we are at the beginning of a new adventure.

Western Christianity is in a process of imagination because of unsustainable decline. What makes church, church? Was churchgoing just a social habit? For many, the ugly yet profound passion story, the guts of Christianity, was avoided. Good Friday is "off" Sunday so fewer worshippers, especially Protestants, participated in the Good Friday religious ceremonies. When death and loss came to family life, the access to Jesus, who had his own terrible experiences, was harder without knowing these things. Holy Week was limited to Palm Sunday praises for Jesus, riding on a donkey (what is not to like in a happy parade?), and the triumph and confidence of Easter stories of the empty tomb and shouts of "He is risen!" Suburban Christianity has an inherent addiction to comfortable pews.

I think the avoidance of the cross was there from the beginning. Who wants to go through that? Yet it is this enigma, rooted in human experience, that made the difference. How can God die? What difference does it make? It was religious imagination that made sense of this tragedy and set people free to counter evil and death with love and courage. Paul was right to preach Christ crucified as the focus for the whole message and ministry of Jesus of Nazareth (1 Cor 1:23). It remains critical for Jesus Green.

Secular storytellers agree. Films of the life of Jesus face the challenge of how to portray miracles and the ultimate one of resurrection. But they all make crucifixion a climax.

These secondary "Gospels" have become more important in our Western cultures where the visual reigns. Far more have seen the films than read a Gospel. The reference points for understanding Jesus have changed. The "little man on the cross" must be understood afresh, and the films of the life of the Jesus are key to making Jesus known, even if they are often at odds with the evidence!

The Gospels are like films: the editors act as a director making crucial cuts in material. Films warn us to avoid idolizing any Gospel and admit our bias. My favorite Jesus film reflects the way I prefer to understand him, met today: *Jesus of Montreal* by Denys Arcand (1989).

Actor Mel Gibson combined his conservative faith with his roles in war movies to direct a film based on John's Gospel.[1] The use of Aramaic in the dialog was meant to heighten the realism of the film, and the physical agony of Jesus came out much stronger. It is true that John's Gospel emphasizes the passion of Christ, but it is a symbolic Gospel. It always provokes a search for meaning that is only made through human imagination. You cannot see, and you would not have seen, the meaning of the crucifixion of Jesus. His death was like the others. You had to know his story to see more.

Consider the subject for which you are the undisputed world expert: yourself. The storyline of your life as a simple series of events misses the fun and fullness of your life. The meaning of life rests on qualities, emotions, the music that stirs, the courage of decisions and commitments. The passion of Jesus speaks to our story because of the need for meaning in the face of our mortality and of those around us. We turn over the paradoxes and contradictions in the Jesus story for our own. Imagination is key. It has always worked that way.

While "John" wrote his Gospel of the Word made flesh, another John, of Patmos, wrote about the end of time or apocalypse (Rev 1:9). This gave us one of the greatest sources for Christian religious art, alongside Genesis and the garden of Eden. The coded religious language permitted John to attack the Roman Empire: the beast that had crucified Jesus and continued atrocious executions of anyone who questioned its power. Revelation warns that faith can be abused and become the opposite of the love it is based upon. It offers a relentless sacred history, to win out over injustice and violence. This imagination is not to escape the frightening challenges of the times but a resource for Christians to find courage and act differently.

When contemporary Christian belief construes the images of Revelation as real, rather than symbolic, it is an attack on this faithful imagination. It takes away each person's responsibility to interpret. This gives me a déjà vu of how Jesus confronted misuse of the law in his time. Today's leaders, who should know better, can impose a mistaken interpretation on others and create a culture of fear rather than love.

If John expected us to trust our imagination, what does this mean? Interpretation and being imaginative are reliable human attributes. Imagination can describe the realities that give us meaning.

You can find meaning and courage with the Gospel of John and the book of Revelation. We do not speak ancient Greek nor use the same concepts. Something else gives continuity: our existence, being alive. This is how we can still get thrills in placing ourselves in somewhere as distant

---

1. Gibson, *Passion of the Christ*.

as the book of Job, the story of a virtuous man suffering for no reason. Religious writing trusts imagination because existential realities are known across cultures, languages, and time.

Both Gospel and Apocalypse assume you can think and believe through symbols. They trust religious imagination as a way of thinking in full color, every dimension, and more. Ask the astrophysicist, the anthropologist, the child seeing her first rainbow . . .

The case to rename Jesus is therefore not so strange. Jesus Christ can be Jesus Green. I stand in the biblical tradition to trust you have the capacity to consider a new name and the meanings it brings. In the film *Jesus of Montreal*, a contemporary resurrection of Jesus comes through the new life of transplant surgery.[2]

I want to reimagine the crucifixion. Jesus Green is seen there best. If not, Jesus Green is a frivolous claim and amounts to sacrilege. This is something that has emerged for me, as it must for you. There is a process that my consciousness relies on. I name Jesus "Green" because of the present crisis of environmental destruction and how it echoes with the remembered drama of the crucifixion: the tension between historical fact and human interpretation. We are confronted with lethal global human impacts and a savior on a cross. This is the power of an event to deal with the fundamentals of existence and the ceaseless stream of living in the now. Imagination is the oxygen of our courage to be.

REVISION IS A MUST

Look at anything long enough, and you will see God. But try literally staring at an object, and your eyes will hurt. Have you discovered holograph printed images that can reveal a 3D outline, only if you look through the image, not at it? The effort to see what lies beyond, pays off. It is a neat illustration for the revision of the cross, as the way to know Jesus Green.

Much of this book has been deconstructing religious ideas: Jesus is further away, as we question the assumptions given to us about Jesus Christ. But I want to recognize the force of Christian faith, nevertheless. We are alive at a time of the collapse of the church as I knew it but not because "that which I call God" has gone away.

Rather, something new is happening.

On January 25, 1959, Pope John XXIII surprised the world when he called the second-ever papal council, Vatican II. I like the Italian description of this *aggiornamento*: an updating, and John's description of it as "opening a

2. Arcand, *Jesus of Montreal*.

window onto the church."³ There was a wave of books and media events that brought the questions and scholarship of universities to popular awareness. A notable book was *Honest to God*, by John T. Robinson. Newspapers made headlines to shock people that there was the "death of God." Ironical really, as the questioning was driven by Christians who sought to make sense of their faith in a living God, well captured in that title: *Honest!* It was the passionate experience of Christians who cried out for better answers than offered in catechisms.

Patterns repeat themselves: there is a gap between the questions and insights of eco-theological scholars and the focus of most sermons. We face an environmental crisis that is bigger than religion, and it exposes the inadequacies of "salvation" as an exclusively human experience. As if this was ever possible, or biblical.

The good news is that people do "mind the gap!" I find talk of a Green Christianity is readily received in worship because the wider world asks these questions. Pope Francis published his second encyclical on May 24, 2015 *Laudato Si'* (Praise be to you), to encourage Roman Catholics to recognize the value of every creature and to question the rampant exploitation of natural resources. It is a fine document, but history tells us not to expect adequate change from the center of power, even when the pope seeks it. Jesus knew this too. I may find Jesus difficult to describe historically, but the Spirit of Jesus is very real, through the stories he told and then others told in turn. I meet the Spirit of Jesus as the ground of my being in human terms, and this Spirit seeks more radical change than churches expect. As ever . . .

At the end of Matthew's Gospel Jesus sends out his disciples with the mission to make followers of all. The Greek is *panta ethne*, all peoples, from which we get ethnic and ethnicity. If Matthew was sure that the message of Jesus was for the whole of humanity, not just Jews, but Samaritans, Ethiopians, Egyptians . . . would he not urge this message serve the whole of the *living* world? Is not the universal good news, news for the whole of the planet? How can it be otherwise, yet this demands a revision of the story so far.

Just as with the "death of God" era, the intuition of what this "Greening" of faith might mean is already felt. Movements have begun within church networks and beyond, with GreenFaith, *Laudate Si'*, and A Rocha groups. Often this takes an ecumenical and interfaith approach that reflects Matthew's *panta ethne* in a hopeful form. I experience this hope directly, with the Greening of our patterns of worship and our sanctuary. Such light, love, and energy has nurtured this book. But it is not enough.

3. Pilario, "Opening the Windows."

Christianity is still a destructive influence in humdrum ways of control and exclusion. I hear of small-time hurts caused by priests who abuse their responsibilities, with antigay statements at baptisms or refusals to respect other religions. The three-hundred-and-fifty-year wait for an apology to be given to Galileo illustrates the moral inertia of a religion that has an attachment to control.[4] This does not limit itself to the churches of empires. I have seen the abuse of religious sentiments by evangelical preachers in Brazil where Jesus was proclaimed as a stage show for public exorcisms. "Come on down!"—Have a moment of being special, despite the dead-end circumstances of your daily life. Fair enough, but do not ask a poor woman to give the little she has to the cause, because she will be back next week for the same fix, to keep going. There was a temple incident with Jesus and a poor widow (Mark 12:41–44; Luke 21:1–4).

The story of Galileo is fascinating. The details reveal it was not a simple conflict between dogma and reason. The delay to put things right, however, is a give-away. This is in part a cultural problem. Christianity has stone churches to express the reliability we seek, to cope with the uncertainties of our lives. Who in the West does not draw a church in their sketch of village life? The stable architecture is also reflected in how worship tames time: we have a routine, a pattern to worship that follows a sacred calendar. Those who do not, like evangelical free churches, have their own beat nevertheless, with endless "fresh" mission goals and strategies.

The sacred time of the church year is cyclical and measured: time as the sequence of one event after another. The New Testament calls this *chronos*. In Greek myth the story of an eternal punishment condemns Sisyphus to push a boulder up a hill, in an endless loop: the striving of *chronos* risks being absurd. But there is another biblical word for time, which is more biological: *kairos*. This is a time when many things come together, like the ripeness of a fruit. We live in a *kairos* era for the transition to sustainability.

I took this sense of good timing to name a spiritual and social well-being charity I founded in 1996, in London: "Kairos in Soho, for lesbians, gay men and their friends." I am glad that the list would be longer today, and tricky to compose as a tagline: for transgendered, queer, intersex, two-spirited and more . . . *Kairos*, because the possibility and the needs were there and then. The vision was for a social and spiritual center to offer an alternative meeting place to the commercial clubs and bars. Whether in Montreal or London, I am told Kairos is needed more than ever, with the collapse of queer village life accelerated by online dating apps. All of us can

---

4. Cowell, "After 350 Years"; Speller, *Galileo's Inquisition Trial Revisited*, 55–56.

be superficial and, in the dating game, swipe to the next picture in the search for the perfect adventure. Easy come, easy go, but where is the depth?

Kairos in Soho called for fruition. After five years of knocking, the doors to manifest this vision did not open; it was not time for a center in the nineties. But we were in the right place and time to respond to a homophobic bombing of the Admiral Duncan pub, next door to our meeting place. Limited, but life-changing resources were made available to us, to offer short-term post-traumatic care to hundreds of people. This alone made Kairos in Soho timely.

Twenty-five years on, and I recognize the same processes of social needs and spiritual insights coming together.

The collapse of church as the center of community life makes it easier to recognize *kairos*, rather than routines of *chronos*. I notice how appropriate it is to use verbs of vision for time. What we see is affected by the quality of time. Kairos is the time of emergence. It is biological time, organic, ripe with change.

Kairos helps a revision of the cross that is critical for Jesus Green. It asks what is going on in the present moment. Theologian Paul Tillich wrote eloquently of seeing this way. He insisted on the nature of human sight to see beyond itself. We see in intricate details, but we are not just photoreceptors like film in a camera; we inevitably see what may lie beneath. The face that is beautiful can also be menacing. An ugly face can hide grace. Sight searches for depth. *Depth offers recreation*. No coincidence that Tillich's writing on sight is found in his book *The New Being* (1955).

Things are emerging. We live in *kairos*. We see depth. We seek a new being for humanity. Suppose that I see Christ crucified in the orangutan clinging to the last tree of her forest. Suppose I weep at this. Suppose you do.

Jesus Green is an emergent truth we share. It is to look at the cross of Christ; to look until you see. A place that was the contradiction of life became the life of Christian faith, to see life for all. Revision is a must, for new things have come together. Honest to God!

FIRST THE TREE

When I moved to Montreal, aged thirty-nine, I joined a water polo team. It was a brilliant way to imagine I was aged eighteen again, because I had first tried it as a student, among the Liverpool University swim team. Then I was way out of my depth. This time I made friends, and I knew it would help me to learn French. What I never expected was the illustration it gives me for Jesus Green, because I am left-handed.

Most lefties, especially those in team sports, know the Latin for left and right. We looked it up because, subtly or otherwise, people treated us differently. Try to call for the ball when you are left-handed! It is a natural bias to look to and pass to the people who catch like we do. Only the best teams feed left-handers. Left- and right-handedness can be serious in some cultures; the left hand is associated with dirtiness and not used in eating or passing dishes at a meal table. We find this bias in language. In Latin, right is *dextra*, and we think of being dexterous (literally handy), and left is *sinistra*, with stories of left-handed people being hit on the hand or worse.

The bias between left and right may have been lost in translation, but now you know, please try to keep your eyes wide open (not shut) when playing sports.

Lost in translation applies to the cross of the Gospels. In Latin, *crux, cruxus, cruce*, in the Vulgate, is similar; but something was lost in the first translation of "cross," from the original Greek to Latin, as Christianity became the religion of the empire. The Greek is surprising. Two Greek words were translated to *crux* in the Gospels, neither mean cross shaped. Both suggest my image of Jesus crucified can be redrawn.

The Uffizi art gallery in Florence shows crucifixions painted through the thirteenth to eighteenth centuries. The clothing and settings reflect the era of the painter, perhaps to bring the cross to that present day or because of limited historical awareness. Still, I assumed the basics were correct. Jesus was nailed to a cross of two beams, a vertical and a horizontal.

Stauros is one of the Greek words for cross, meaning rod or stake. Only an upright rod was needed to kill someone. A stand of stakes would mark out a location, like Golgotha, as a permanent warning. The word "stakeout," referring to police surveyance, still carries this voyeuristic function to the execution.

If the upright rod of the cross was already in place, ready for the next victim, and there may not have been a horizontal beam, then this shakes up how I see the passion story, in a good way. In 1968, archaeologist Vassilios Tzaferis excavated a Jerusalem tomb that contained the bones of a crucified man named Yehohanan.[5] This gave new details to the norms of Roman crucifixion methods. It confirmed ancient literary sources that describe crucifixions quite differently to those we have seen in films, or with today's Easter reenactments by Philippine Christian communities. Feet would be nailed together through their sides, a matter of practicality to save metal and effort: one nail.

---

5. Tzaferis, "Jewish Tombs."

Crucifixion with a single stake was drawn by a renaissance scholar Justus Lipsius (1547–1606). In his book *De cruce* (1594) Lipsius describes the variations in crucifixions, including "vertical" executions: *cruxus simplex ad affixionem*. This recalls the many images of Saint Sebastien: a soldier saint executed by being tied to a stake or tree and shot by arrows.

Crux Simplex (1594), by Justus Lipsius (1547–1606); *St. Sebastien* (1525), by Sodoma or Giovanni Antonio Bazzi (1477–1549), Uffizi Gallery, Florence

Lipsius drew on Roman descriptions that suggest there was a routine to this method. Was Jesus' execution routine? To change the routine would express something about the victim or the practicalities of execution that day. This history at least raises questions.

For a cross-shaped execution victims could be tied rather than nailed at their wrists, after being forced to carry the horizontal crossbeam, or patibulum. That Jesus carries a cross in the Gospels suggests the usual cross-shaped images we have in our heads. If there were two thieves either side of Jesus, still alive after three hours, this matches too; a crux simplex killed more quickly. In all cases, however, execution was through hanging the victim on a stake. If a main stake was left in place this saved time and timber. There is evidence this was more likely, given the descriptions of the Jewish and Roman historian Flavius Josephus, thirty years after Jesus' execution:

And now the Romans, although they were greatly distressed in getting together their materials, raised their banks in one and twenty days; after they had cut down all the trees that were in the country that adjoined to the city: and that for ninety furlongs round about; as I have already related. And truly the very view itself of the country was a melancholy thing. For those places which were before adorned with trees, and pleasant gardens, were now become a desolate country every way; and its trees were all cut down. Nor could any foreigner that had formerly seen Judea, and the most beautiful suburbs of the city, and now saw it, as a desert; but lament and mourn sadly at so great a change. For the war had laid all the signs of beauty quite waste. Nor if any one that had known the place before, had come on a sudden to it now, would he have known it again: but though he were at the city itself, yet would he have enquired for it notwithstanding.[6]

Josephus also describes how this lack of wood arose through sieges, as with Pompey's siege in 63 BCE:

For on the parts towards the city were precipices; and the bridge on which Pompey had gotten in was broken down: however, a bank was raised day by day, with a great deal of labor; while the Romans cut down materials for it from the places round about.[7]

These accounts tell us there were trees to be felled, so many that the war changed the landscape drastically. But not enough to meet the Romans' needs. Trees were already precious and timber for crucifixion limited and reused. It is likely a well-established practice.

The book of Deuteronomy, seventh century BCE, urges trees be saved:

When you besiege a city you shall not destroy its trees; for is the tree of the field a man, that it be besieged by you? (Deut 20:19)

The tree, now a stake, came first. Jesus was killed at the stake, by spear or suffocation.

The other New Testament word for cross is *xulon*; it too provokes a rethink. *Xulon* means wood or a tree itself. There is a sort of honoring of the tree in this double meaning, as it remembers wood had life to it. (Do you eat lamb?) This is explicit in one saying of Luke's Gospel: "For if they do these things when a tree [*xulon*] is green, what will happen when it is

6. *Wars*, 6.1.1, in Josephus, *Complete Works*, 872.
7. *Antiquities*, 14.4.2, in Josephus, *Complete Works*, 443–44.

dry?" (Luke 23:31). Luke recognizes the cross as first a tree and then wood. He chooses *stauros* for other references but goes back to *xulon* in his sequel book, the Acts of the Apostles, with the speech of Peter to the astonished Jews gathered in Jerusalem for feasting: "The God of our ancestors raised up Jesus, whom you had killed by hanging him on a tree [*xulon*]" (Acts 5:30).

This sense of corruption of the tree was already associated with the practice of crucifixion (Deut 21:22–23), and Paul's Letter to Galatians quotes it: "Christ redeemed us from the curse of the law by becoming a curse for us—for it is written, 'Cursed is everyone who hangs on a tree'" (Gal 3:13).

A third author, claiming to be Peter, wrote between the time of Paul and Luke: "He himself bore our sins in his body on the tree [*xulon*] so that, free from sins, we might live for righteousness" (1 Pet 2:24).

Trees are important to the big-picture story of the Bible, in the first and last books. Genesis names the tree of life and the tree of the knowledge of good and evil in the garden of Eden. Revelation describes the tree of life in the new Jerusalem of a new heaven and new earth; the leaves of this ultimate tree are for the healing of the nations. Trees are like the bookends for the story of the people of God.

Christian artists of the early Middle Ages let their imagination inspire them to paint the cross as a tree of life. The earliest English literature captures this too, with the religious verse "A Dream of the Rood [Cross]" (c. 400 CE). Rood, or rod, is from *stauros*, and the poem offers the tree's point of view.

> Till I heard in dream   how the Cross addressed me,
> Of all woods worthiest,   speaking these words:
> "Long years ago   (will yet I remember)
> They hewed me down   on the edge of the holt,
> Severed my trunk;   strong foemen took me,
> For a spectacle wrought me,   a gallows for rogues.
> High on their shoulders   they bore me to hill top,
> Fastened me firmly,   an army of foes!
>
> "Then I saw the King   of all mankind
> In brave mood hasting   to mount upon me."[8]

The "Viking-era" notions of the kingship of Jesus overtake the anguish of Gethsemane, yet these words take fresh meanings for our times:

---

8. "A Dream of the Cross," in *Earliest English Christian Poetry: Translated into Alliterative Verse, with Critical Commentary*, by Charles W. Kennedy, 93. © 1952 Hollis & Carter. Reproduced with permissions of the Licensor through PLSclear.

"With black nails driven
Those sinners pierced me; the prints are clear,
The open wounds. I dared injure none.
They mocked us both. I was wet with blood
From the Hero's side when He sent forth his spirit.

"Many a bale I bore on the hill-side
Seeing the Lord in agony outstretched . . .
Wan under heaven; all creation wept
Bewailing the King's death. Christ was on the Cross.[9]

Life itself, not life in the shadow of an eco-driven catastrophe, inspired this poet to find the tree and Christ mocked together. A nature-Jesus dialog provoked by the memories of his crucifixion is not new. But this dialog now has unprecedented urgency.

Trees. If we had to chose one form of life besides human beings, perhaps trees would suit, to be the "other" of our existence. As we tell a scientific story of life on earth, should we not revere trees, as indigenous people already do? Trees were before Homo sapiens, before the first dinosaurs. Trees made possible the ramification of life, like a river delta, out of which Homo sapiens emerged. Yet trees are taken for granted. Not seen. The Gospels tell of Jesus being crucified, not the tree being felled. Why should it be otherwise until now?

Now we notice trees; they are part of the story of Jesus, as Zacchaeus climbs; mustard seed grows; birds make nests; Judas hangs. Now we notice the death of the tree; someone chose it and fashioned it. The finale of one of the most important stories ever told, made possible by wood. We notice what has always been there. The tree waited. "You can do this," it says, "You can use me to make this hill a place of death."

In Lent 2023, I took a potted pine tree to encourage my congregation to follow the way of the tree, and a candle to follow the Christ. Each moved, nearer to the front of the sanctuary, week to week, toward the fatal meeting of Good Friday.

Was it an oak, a walnut tree? Where did it grow, and how old was it when felled? Think on the wooden cross. To help, let me tell you a story from another place of death.

Hiroshima, Japan: at 8:45 a.m., on August 6, 1945, and seventy-five years to the month as I write, an atomic bomb destroyed Hiroshima. Over

---

9. "A Dream of the Cross," in *Earliest English Christian Poetry: Translated into Alliterative Verse, with Critical Commentary*, by Charles W. Kennedy, 94. © 1952 Hollis & Carter. Reproduced with permissions of the Licensor through PLSclear.

eighty thousand people died in an instant. The photos taken of it are haunting. Survivors recalled that everything in the epicenter was black, white, or grey. The rumor no life would return for decades was believable, and who would live to see it? The remaining trees shaped the memories of survivors: so much else was flat. They gave some reference point and limited shelter. Then they proved predictions wrong. All that was left of these trees were their vertical trunks and remnants of branches, but these budded and burst open. The leaves gave new meaning: life would be restored.[10]

Tomoko Watanabe, the cofounder of Green Legacy Hiroshima (GLH), was born after the war and raised in Hiroshima. She would play in the park unaware of their extraordinary story until her friend, cofounder Nassrine Azimi, pointed them out. "The trees taught me many things," said Watanabe, "I began to love them and wanted to tell other people and the next generation about them."[11]

Now you can apply to GLH, as my church has, along with three Westmount communities, to receive the seeds of survivor trees. We have a trail of *Ginkgo biloba*, to see and touch life that survived devastation, to revision peace, through nature's resilience. Every spring I have a *frisson* of this terrible experience as I wait for signs of new life with buds and leaves. It is also an Advent sign, when I imagine of the coming of that tree of life in Revelation, to heal the nations (Rev 22:2).

The tree was first, then the stake: *xulon* became *stauros*. Human beings instrumentalized the tree. Does the tree leave the stage because the story moves to an empty tomb? Not so easily, because in real life we are not at resurrection but crucifixion; we are in *chronos* looking for *kairos*: an escape from human failures. More: the "overshoot" of humanity demands new energy and movements, choices and consciousness. God forgive us, "for we know what we have done."

Does the tree have a way to speak as never before, because we have taken the cross for granted? The tree helps us see the cross deeply. It waits for us to go through the Jesus story, to discover how.

OLD JESUS GREEN

I would not be Christian if the past meant "gone and buried." Old ideas can be like old buildings: they fade away with neglect or scintillate if cared for. The impeccable state of the White Tower of London astounds me, compared to other castles, because it has been occupied since its construction in 1097.

---

10. Hiroshima for Global Peace, "Meeting Trees."
11. BBC News, *Our World*, 1:48 a.m. BST.

The story of Jesus relies on the care it receives from present followers. Given we are, as ever, a ragtag band, this can go as wildly as a Pollock painting, but the Bible anchors us to pay attention to what was written, just as you keep your eyes on the road to drive.

Therefore, the meaning we find in Jesus Green must be latent in Jesus of Nazareth, at least in the meanings people took from him for their journeys. Those meanings claim a connection between you, me, Jesus, the living world, and the existence of the universe.

Millions of people were crucified, but name any other, besides Jesus. Perhaps Spartacus? It seems the punishment was effective in removing people from history. For Jesus it was a new beginning. People remembered he refused titles and public proclamations when he was alive. Once dead and risen, names and titles were essential. Who is this person you are so joyful to tell me about? Why would this happen to him and his intimate followers? The titles for Jesus try to make sense of his story on an extraordinary scale.

Three of the names given to Jesus draw him to our Green concerns. Jesus is divine wisdom; son of man; Word made flesh. History tells us the claims for these names were far from the detached discussions you may have had at school or in the bar.

In my church sanctuary, the chancel ceiling is painted with symbols of the twelve tribes of Israel and the twelve apostles. Many of the apostles' symbols recall their early death in different forms. The early Christians faced rejection and misunderstanding that jeopardized their lives. As a small Jewish sect that lived through the destruction of the temple in 70 CE, they were not assimilated within the Pharisees, who produced today's rabbinical traditions. But a new religious movement was vulnerable. Why take these risks? I do not believe it was to die young like Jesus, anymore than Jesus wished to die, but rather a reflection of the existential change that Jesus brought to those who truly knew him. They found he taught with the authority of the God they believed in, to offer them new hope and meaning. This proved priceless, like the pearl in a parable, and it revealed God to be a God of love.

For us, the crucifixion grounds the story of Jesus in historical realities. It confounds us, unlike any other detail, and points to the socially challenging nature of Jesus' message: you would get crucified for that.

The earliest Christian writings, the letters or epistles, make amazing claims for this carpenter's son, some twenty years after his death. Jesus is called the Messiah, Christ (the anointed one), not simply rabbi, a teacher. "Lord Jesus Christ" is a consistent full title, such as: "To the church of the Thessalonians in God the Father and the Lord Jesus Christ" (1 Thess 1:1).

These titles claim Jesus fulfilled the promises of God to send a savior, anointed like a king or queen, with power to bring the realm of God to everyone in the foreclosing of life on earth.

Our trouble is that we are so exposed to this language to describe Jesus that we fail to grasp its novelty and intent. Neither have we been prepared to look out for a Messiah unless we are Jewish. Jesus knew the risks of being claimed Christ when he attracted crowds for healing and teaching, but he avoided them. The crucifixion took away his control. It became the moment people looked back to, for answers. Jesus' identity and death are inseparable.

The Gospels of Matthew and Mark suggest it was the manner of Jesus' dying that made him stand out. But I agree with today's scholars and vulgarizers: it was the continuity of Jesus' healings, teachings, confrontations, and prophecy, as well as his cross, that informed the experiences of his resurrection and the meanings people made of them. These describe his nature, not his appearance, beyond wounds. People found him alive and recognizable, in their minds and hearts, with dreams and visions. A better word for this than continuity is integrity.

If I claim Jesus is integral to life, it makes more sense than a debate on him being the second person in the Holy Trinity. Jesus' integrity made love known beyond death and made his followers courageous too. They had found something beyond life itself. The process to name Jesus was historical, not supernatural, as it is with Jesus Green.

The names for Jesus were offered while Christians belonged to marginal communities without expectations of grandeur or social control. They were attempts to validate Jesus within a larger divine drama. If he had been seen as the Messiah, and his relative John the Baptist as Elijah returned, his death and resurrection turned things upside down. The real Messiah would not have taken that path. Does this explain why the Greek title Christ was preferred over the Jewish Messiah? Perhaps.

Christians from the beginning have put the cross first, as essential to his Way; the passion of Jesus was the cauldron of a new religion. It hinged on paradox: how a wasteful, ineffective death brought out courage and new communities. This paradox nailed down Jesus' nature as a wisdom to bring new power into human history. He had shown it in his teaching and choices. The risen Messiah has wisdom. Jesus is the wisdom of God and the power of vulnerable love.

It may already make sense to you that wisdom deals in paradox. We are on the same page as St. Paul. These words are in the first chapter of one of the earliest New Testament writings, his letter to the church in Corinth:

> For Jews demand signs and Greeks desire wisdom, but we proclaim Christ crucified, a stumbling block to Jews and foolishness to Gentiles, but to those who are the called, both Jews and Greeks, Christ the power of God and the wisdom of God. . . . Christ Jesus, who became for us wisdom from God, and righteousness and sanctification and redemption. (1 Cor 1:22–24, 30)

Corinthian ideas about wisdom were associated with rhetoric and elitism. We still do this: someone is a "wise so-and-so." But Paul claims a different sort of wisdom in these verses. It is "from left field," *sinistra*, ridiculous unless paradoxical, and then it claims more than a risen hero. It suggests, in the first experiences of the risen Jesus, people reconnected with the whole of creation.

Further in his letter, it seems Paul sees Christ as a cosmic force holding existence together.

> Even though there may be so-called gods in heaven or on earth—as in fact there are many gods and many lords—yet for us there is one God, the Father, from whom are all things and for whom we exist, and one Lord, Jesus Christ, through whom are all things and through whom we exist. (1 Cor 8:4–6)

Three different strands of ideas are found in this passage: Jewish belief in one God, Christians following one Lord, and Stoic ideas in the use of the words "from," "through," and "all things" in reference to God. James Dunn finds this exalts the creative role of Christ in the present experience of the Corinthian church and across the world.[12]

Paul insists that the spiritual knowledge of the Corinthians has a material reality. Christians are an extension of creation, as Jesus saves Corinthians in the same manner that makes the rest of the living world and its environment possible. Paul had a strong sense that Christ crucified and risen was in the everyday fabric of life, in all its dimensions.

Where do these claims come from? They are so different from the modest Jesus who rode on a donkey into Jerusalem rather than a horse, or taught people to be self-effacing servants of one another.

The explanation lies in the extraordinary experiences of the risen Jesus. Being raised was equated with being the source of life itself. Only such power could surpass death. Christian conversion resonates with God's act of creation, as resurrected life is a new reality (2 Cor 5:17; Col. 3:10; Eph 4:24). When Paul wrote of this "new being you have in Christ," he had been

---

12. Dunn, *Christology in the Making*, 179–80.

through this himself: he was made blind for days and turned away from the persecution of Christians. Conversion is experienced as a continuation of the creative power of God.

We are back to integrity, and the impression Jesus made, before and after his crucifixion. The stories of Jesus who unmasked religious hypocrisy, refuted challenges to his authority, and told subversive parables relied upon a wisdom Paul proclaimed. Resurrection power is the power and wisdom of creation. When I urge us to consider the cross, to name Jesus "Jesus Green," this is faithful to Paul.

The fullest description of Jesus as creative wisdom is found in the letter to the church in Colossae:

> He is the image of the invisible God, the firstborn of all creation; for in him all things in heaven and on earth were created, things visible and invisible, whether thrones or dominions or rulers or powers—all things have been created through him and for him. He himself is before all things, and in him all things hold together. He is the head of the body, the church; he is the beginning, the firstborn from the dead, so that he might come to have first place in everything. For in him all the fullness of God was pleased to dwell, and through him God was pleased to reconcile to himself all things, whether on earth or in heaven, by making peace through the blood of his cross. And you who were once estranged and hostile in mind, doing evil deeds, he has now reconciled in his fleshly body through death, so as to present you holy and blameless and irreproachable before him—provided that you continue securely established and steadfast in the faith, without shifting from the hope promised by the gospel that you heard, which has been proclaimed to every creature under heaven. I, Paul, became a servant of this gospel. (Col 1:15–23)

These verses mean to cover everything. They show how Jesus operated as an icon, not an idol. He is not God, but God can be known through his image. They claim life comes from a prime source; that power and power systems exist by the same creative birth, and ultimately the contradictions of evil and failure will be reconciled to God.

The later Jewish-Christian letter of Hebrews opens with a cosmic declaration:

> Long ago God spoke to our ancestors in many and various ways by the prophets, but in these last days he has spoken to us by a Son, whom he appointed heir of all things, through whom he also created the worlds. He is the reflection of God's glory and

the exact imprint of God's very being, and he sustains all things by his powerful word. (Heb 1:1–3a)

Much of early Christian writing dealt with guilt and forgiveness, but we are going further than finding forgiveness here. It declares the creative, sustaining Jesus who spoke for God and reflected God's glory. Paul's best-known description of what this imprint may be is the famous "faith, hope, and love" of 1 Cor 13, and yes, it is good to marry by!

The heretical claims of Christians about Jesus' identity did not come out of thin air, but from Jewish and Stoic philosophies as they sought to explain Jesus to their neighbors. This made Jesus wise before he was born, and this wisdom is manifested in creation. Perhaps creative wisdom is more appealing today than the religious title of Christ; wisdom is shared across religions and found in deeper Green thinking. Jesus noted the diversity of life that one tree offers as a source of wisdom for his teachings, and he suggested we are to go further (John 14:12). He invited people to name him, rather than announce himself (Luke 9:20).

"Jesus, the son of man" is another old name for Jesus that encourages a new one. But this is a tricky title. Scholars are still arguing about it because it has different meanings according to its context, like "she is such an artist" could be an acclamation or a put-down. "Son of man" could mean someone like you or a representative of humanity, and maybe a special one at that. In the Gospels it is limited to the sayings of Jesus, and many think Jesus used it to talk about himself. Hence the debate.

As the Christian story spread, it risked distortions all the time, so second- and third-century Christians were forced to wrestle out a definition of their Lord. They took the two titles "son of man" and "Son of God" to have equal importance: Jesus was defined as fully human and fully divine, not a temporary human being perfected by God nor simply God descended in human form. This pairing rather hid the rich ambiguity of "son of man," and the scholars who uncovered this have arguably brought the historical Jesus a little closer. Why would "son of man" be chosen if it had ambiguity, when the church wanted clarity, unless it had authority as original to Jesus? When he called himself "a human being like you," Jesus expressed his solidarity. It encouraged people to expect goodness within themselves if they saw goodness in Jesus. Just as he taught, "the realm of God is among you" (Luke 17:21).

I find this resonates for us with Jesus Green. Of all the titles given to Jesus of Nazareth, it was that "son of man," a human being, someone like you, who was arrested and crucified. Then it can flip: this Jesus was also *not* like you or me. Who has his courage? Yet potentially, you could be like him, he says, if you are willing to serve your neighbor (Matt 12:25–28). He disturbed ordinary people to live beyond their fears.

The fact that "son of man" arguably goes back to Jesus strengthens our revision of the cross: to see all of humanity in one person, meeting the nature it has killed on the cross. At the least, it helps me to trust Jesus would smile in being called Green.

The Gospel of John gives us a third title for an "old" Jesus Green but not without risk:

> In the beginning was the Word, and the Word was with God, and the Word was God. He was in the beginning with God. All things came into being through him, and without him not one thing came into being.... And the Word became flesh and lived among us, and we have seen his glory, the glory as of a father's only son, full of grace and truth. (John 1:1–3, 14)

This profound passage can grab us despite our twenty-first-century mindset. But there is a duality to John's Gospel that unfortunately became a trend in church teachings, to downgrade the material in favor of the spiritual. This meant the cosmic glory of this opening, as an affirmation of the sacredness of earth, has often lost out to a focus on human souls alone, and is more than an intellectual game if it encourages contemporary Christians to devalue the natural world. It makes Christianity a potential adversary to ecological change. Equally, critics of the church have imagined those of us remaining are stupid or anti-scientific, in worlds apart.

But there are bridges, just as John's prologue bridged between Jewish and Greek thought. When I offered to train for ministry, I was already affected by Paul Tillich's theology, of a God who was existential if anything and everything. We can step away from classic Christian metaphysics and still find God is "real," or is this "reality is God"? One way is to notice the less popular descriptions of Jesus, as we did with "wisdom" and "son of man."

An exaltation of Jesus, rather than a doctrine of incarnation, allows Jesus to be a great man, even somehow the wisdom driving the manifestation of the first human beings, a sort of quintessential human, and the force of nature at the same time. This Jesus confronts us with what it can mean to be a person and shape civilized societies. We can be uncertain about God, but follow the Way of the anointed one.

Whether you are a poet like Walt Whitman or a scientist like Jane Goodall, the power of nature has a wisdom and demands respect. The more one observes and understands, the more the living world leaves us astounded and humble. The complexity of biochemistry, yet the elegance of a DNA double helix; the apparent hierarchy in ecosystems from bacteria to plant, to herbivore, and carnivore has circularity and interdependence; micro-reality reflects in macro-phenomena. That we are, that consciousness exists out of inert matter, continues to be an enigma. Yet there is a sense of flow, cause and effect—the directionality of time, the nature of the universe making this possible. All this, in our humanity.

I have shown my bias to prefer "wisdom" and "son of man" over "the Word," but for those who are diehard theists, the Word made flesh can still be eco-gospel: Sallie McFague is one of the significant theologians to contribute to a more contemporary cosmic Christ, drawing attention to ecology as words about home.[13] She suggests we see the world as the body of God, which is being wounded in the destruction of the natural planet. We change the natural world and create our own habitats, but the challenge is to find a sustainable place in the world. If habitats are arguably a true definition of home, beyond the human, then habitat destruction can be seen as a form of desecration. The cosmic Christian claim of Jesus' incarnation validates more than human flesh, theologically and scientifically. The Word made flesh is geosphere and biosphere combined, "ashes to ashes, dust to dust," around and around, so can we find the molecules of Jesus of Nazareth today?

This deep incarnation corresponds with deep Green thinking and gives Christianity a capacity to articulate for the sacred nature of the earth. In 2017 I called a Good Friday pilgrimage to the threatened Technoparc wetland, on the northeast side of Montréal-Trudeau International Airport, to witness a different sort of crucifixion.[14] Christians have a basis to work with the other traditions that have long called for honoring Spirit and Mother Earth, as much as the secular eco-activists who are there already.

The ancient titles for Jesus suggest he waits for recognition as Jesus Green. Each comes from the life of his passion and claims a profound sense of connection. Jesus' command to "love your enemy" is as sharp as ever, and for all (Matt 5:43–48). It goes deeper than any present knowledge of humanity, and this love is Green; love of others as ourselves celebrates the whole living world.

TWO WHERE THERE WAS ONE

The crucifixion of Jesus is the climax to his ministry. Did you think that? It was the ultimate anticlimax in so many ways. Would the Roman governor accept the Jewish leaders' rejection of Jesus? Would God permit it? Would

---

13. McFague, *Body of God*, 56–57.
14. *Westmount Mag*, "Saving the Technoparc Wetlands."

angels save Jesus? What about some burning bush to stay the hand of executioners?

Nothing stops the tragic end. Even though Jesus had been whipped to a critical condition, hope remained until his last breath. Crucifixion was a shameful, unclean death, and the Gospels put abuse onto the lips of onlookers: "He saved others, now let him save himself!" (Matt 27:42). Yet this is when the work of Jesus is fulfilled. That is the extraordinary Christian belief, from the safer distance of resurrection faith. Fulfillment meant the discovery of a new force to Jesus' teachings, which is present in the gospel of Jesus Green.

> And he said to them, "Therefore every scribe who has been trained for the kingdom of heaven is like the master of a household who brings out of his treasure what is new and what is old." (Matt 13:52)

A stake awaits the victim of Roman justice. The dead tree and the dying savior meet. Two become one, become two. If this seems a riddle, the Gospels tell us Jesus challenged the religious leaders of his day with a riddle to subvert a simple understanding of the Messiah:

> Now while the Pharisees were gathered together, Jesus asked them this question: "What do you think of the Messiah? Whose son is he?" They said to him, "The son of David." He said to them, "How is it then that David by the Spirit calls him Lord, saying,
> 'The Lord said to my Lord,
> "Sit at my right hand,
> until I put your enemies under your feet" '?
> If David thus calls him Lord, how can he be his son?" No one was able to give him an answer, nor from that day did anyone dare to ask him any more questions. (Matt 22:41–45)

Jesus quoted from Ps 110, which was taken to be written by David: the Lord God addressed the Messiah, who is placed at God's right hand. The riddle is solved if the Messiah has honor beyond death, whenever he comes, and this surpasses the biological father-to-son relationship of the "son of David."

Now there can be a riddle of Jesus Green: "If Jesus the Christ is the source of all that is created, the creative wisdom and Word, he is present in the wood of the cross as much as in his human body, so that Jesus is nailed to himself. What sort of purpose is this?" I invite us to a way of "seeing" the crucifixion that is truthful beyond conventional Christian doctrines; something new because new parts are brought to the whole.

The reconciliation of God to humanity and to all of creation, "creation groaning" (Rom 8:22), takes place in a moment of time, in a place located by this particular cross. It waits for us.

If this Jesus is Jesus Green, we see this crucifixion afresh. Or classically, when we see the crucifixion this way, we see Jesus Green. The wooden cross has depth, like Jesus. Species are destroyed because our species breeds, and consumes, on a colossal scale. This makes his cross. We see the death of these species in solidarity with the death of Jesus, rather than causing it. For who, truly, is in control?

The cross is the reality of Gaia, with climate change, viral pandemics, agricultural collapse, and more; we are as Jesus was, caught up in systems, vulnerable to death, because we cannot escape the human-nature reality. We crucify ourselves. Biophilia is wasted, it is the abuse of whom we love, human and nonhuman.

A new paradox emerges: when Jesus of Nazareth was condemned, two had to become one. His execution required a certain level of restraint or union. The cross has been noticed as a special object, even an object that could become a tree again, with a level of imagination and magical thinking. The cross was seen as an ultimate holy relic, so it is said the wood of the holy cross would now build a galleon! Past Christians saw the cross only because Jesus was on it. The focus was on him. But as never before do we see two.

What we see is different: two become one but not completely; we can never lose sight of the differences. We see a vital dissonance of the cross of Jesus Green: a crucified dialog. In place of the loneliness of the cross for Jesus Christ, Jesus Green was never alone: he died into nature, and in the moment of death, "It is finished." The two become one. A fusion. "It is finished" has the sense of completion: integration and revelation, with the effect it may have on us.

Many artists focused on the next step, taking Jesus from the cross, with the grief it entails, when once again the person of Jesus matters most. But stay with the cross and the dialog of an emergent crucifixion: the saving activity of God manifests in our consciousness, calls us to repent, to change again.

You are reading this book because Christianity has not yet articulated a strong enough respect for nature. But Christian transformation has always come from the cross. From the cross of Jesus Green, we can break out of alienation between Christians and nature, humanity and the biosphere. The dialog between nature and humanity names a new crucifixion as horrific as the first. Suddenly, this is not about religion, as it never was. It is beyond a

Jewish teacher and healer, even beyond hopes for a Messiah. It is something of all of us, and something of all of existence, in that awful death pinned upon death.

Jesus Green dies to reconcile the needs of humanity to the needs of nature. It is symbiosis crucified. His death exposes how nature and the best of being human die together. The dialog between humanity and nature, from the cross of Jesus Green, allows me to escape the anthropocentrism, the human-centeredness of Christianity, rightly criticized by ecologists.

Yet overshoot is a human fault. Nature demands us to find ways our species can flourish, without decimating the cycles and systems of life. Our task is to understand ourselves as never before. This is the deep Green work to understand our systems, as well as our individuality, and it resonates with sacred history. It was systems and individuals that crucified Jesus of Nazareth.

The cross of Jesus Green calls us to be courageously self-critical; to look deeper into the truth of our responsibilities, how we live, what we kill.

ONCE AND FOR ALL

Most of us have been to a party that gets out of hand because of drugs and alcohol: foolish things are done, and people get hurt. The morning after, the scene is cold and begs the question, "Was it worth it?"

The crucifixion of Jesus was meant to be the final indulgence of human folly against the Creator, a story of rejection and scapegoating of the ultimate messenger.[15] I was taught that the empty cross celebrates the resurrection of Jesus Christ. Today, I see it as a place of desolation. I see a single stake on the skyline, intentionally left. It states you may be the next victim. It is a place of death.

While the Gospel writers take the story elsewhere with Jesus' body laid in a tomb, this place remains, desolate, real.

I have seen another place, Indonesia, where the remnants of a tropical forest are burned for plantations of palm oil. The crucified is an orangutan clinging to the last tree, her home destroyed. The waste of it appalls. More than the crucifixion of Jesus of Nazareth, this moves me to anger and despair, because the first was meant to prevent the second. It is evil incarnate; crucifixion returned.

Images of the last tree of the burned forest force me to admit the powerlessness of God in human terms: we destroy the beautiful, gentle apes of

---

15. Young, *Construing the Cross*, 43.

the forest because of greed and poverty. Global systems devour people who know they are doing wrong to slash and burn.

Yet my attempts to draw a parallel are possible only because of the one crucifixion: the event of Jesus of Nazareth. The vulnerability of God also comes with a sense of a lodestone that has changed everything. I find new power in the cross, with our contemporary disobedience and sin against the integrity of creation. That empty cross was the ultimate place of death and failure. In classic terms it calls our souls to resist evil and choose good, once and for all.

The power of "once and for all" is a remnant process: the remnant of Christianity, of the natural world, of human nature. Who and what will survive, when we finally wake up and see the party of human indulgence is over?

If you are not religious, thank you for your trust, as I press these points. Perhaps you believed me that this story is not to defend churchgoing. You heard me when I wrote of Jesus and crowds and not seeking control. When and how you survey the cross is your affair. It has always been that way.

Unless you have had a profound spiritual awakening it is difficult to describe what a confrontation with the cross is like. One comparison is how we fall in love. This may be so strong that you are willing to go beyond family and kin's expectations, like Shakespeare's tragedy of *Romeo and Juliet*. But romance may not extend beyond the importance of the beloved. A religious awakening, at least in the great religious traditions, breaks down the grip of ego on the self. Romance is but a tributary to an ocean of reality that makes *everything* full of glory and possibilities, and not for one's own self. There is the sense of extraordinary belonging:

> I am wind on the sea.
> I am a wave of the ocean.
> I am the roar of the sea.
> I am a powerful ox.
> I am a hawk on a cliff.
> I am a dewdrop in sunshine.
> I am . . .
> I am a boar for valour.
> I am a salmon in pools.
> I am a lake in a plain.
> I am the strength of art.
> I am a spear with spoils that wages battle.
> I am a man that shapes fire for a head.[16]

---

16. O'Cléirigh et al., *Leabhar Gabhála*, 263–65.

I took the tragedy of orangutans with a purpose. What is different to earlier "visions" of the cross with Jesus Green is the freedom to be nonreligious. The vision of the cross is reflected as passion, as well as beauty, in the eye of the beholder. A better-known African American spiritual makes this journey of imagination explicit:

> Were you there when they crucified my Lord?
> Were you there when they crucified my Lord?
> Oh, sometimes it causes me to tremble, tremble, tremble.
> Were you there when they crucified my Lord?

"Were You There" was first published in William Barton's *Old Plantation Hymns* (1899), but it predates the American Civil War as a slave song.[17] The followers of Jesus found they were part of the problem: "Am I that unforgiving debtor? Do I build my life on sand or on rock? Am I the priest who crosses to the other side of the road?" The African American experience read the cross, to the point of trembling. Slavery, a huge theme in the Bible, demanded hope of resurrection for those it had killed, along with the bringing down of those served by it. It offered meaning in the face of hopelessness, evil, and death, not to accept present crucifixions but to insist that Jesus was crucified "once and for all." The future can be different, through a cross that makes us tremble.

I have never seen a crucifixion, and would not look on its brutality, but I have had the privilege to lead many funerals. Each is unique, despite similar words and actions. I never know how friends and family will react, especially when I am a guest officiant.

An elderly matriarch of a large family died, and I was present as people gathered before the ceremony to pay her respect. Dignity and kindness flowed in this, and the service itself. Her burial was within sight of the chapel, on an uplifting summer day. Yet as her coffin left our sight, several were overcome with anguish. It was as if this was the moment of her death: somehow up to then, the reality had not gone in. I remembered the story of Jesus who cried with the family of Lazarus: "How much he loved him" (John 11:36). I had been given privileged access to love that could not let go. In that grief, I was welcomed into the life force of this woman and the love of the lives she cocreated, and even remotely, I became part and share this beauty with you. Ordinary is extraordinary.

I find echoes of the Spirit of Jesus in this matriarchal story. Who died on the cross? Jesus was symbol and sage. The Christians who faced his shameful death found their grief met, then surpassed: they found Jesus

---

17. Glover, *Hymnal 1982 Companion*, 349.

embodied the source of life, who answered the fact of all deaths, not just yours or mine. Jesus had spoken words that were maternal, creative, sustaining. He gave up his Spirit in ways that shook the earth. This was beyond a human death, or even the death of humanity. That is the impression that we have received, which would make us tremble. He had taken people beyond the human, with his love of nature, his "Father's" love, that reaches even sparrows (Matt 10:29).

Those are the ancient claims with truth for our twenty-first-century salvation.

We see a cross of two, where there was once one. Life in dialog, not death alone.

Once and for all, because once you know it, you will always find the same truth, in each death, in each life, as the gospel of Jesus Green: home for all, not just for humans.

Chapter 7

How Green Is Our Savior

THE WAY AS THE ANSWER

Dread for our planet is real. It comes with guilt and begs for antidotes. For many, the scale of the human destruction of nature takes them to despair, apathy, or avoidance.

In very different times, when Jesus resisted easy answers, despite the despair of crowds, he posed new questions: "How could the poor be blessed? Why not hate our enemies?" He challenged platitudes, and John's Gospel claims Jesus described himself as the Way. The same Gospel has Pilate ask the question, "What is truth?"

I have tried to prepare a Way and share truth without misrepresenting a tradition that has done harm as well as good deeds. Now I try to combine the elements that have emerged through this Jesus, as a paradoxical, specific contribution to sustainable civilizations. This is to insist that spiritual and scientific understandings work together to offer solutions, as never before.

How can we change? How can humanity reduce and reorder its demands on the planet? How can we contribute to the health of the living world? Does a model of "life out of death" fit science and faith because the ego of humanity as a separate entity must die, for life to flourish again? I invite the discovery of faith with Jesus Green, in your terms. It is not just wordplay, but substantial, to find what can make common agreements: the principles for life as the "top" species at a time of crucifixion. I invite you not to look away but to follow as best you can, to discover how the particularity of Jesus Green resonates with life itself.

LIFE WITHIN LIMITS

How many times a day do we get the message that limitless freedom is happiness and success? Like a car on an open road, a luxurious retirement, the worry-free existence, more cash than you know what to do with . . . Yet a little knowledge about life tells us it is the very limits, the rules and relationships within life, that mean we are here at all.

Science and the Bible agree that life appeared from the organization of matter on an epic scale. A good planetarium astounds me with how our solar system came into being, a rudimentary planet Earth and some physical stability on its cooling surface. The chaos of high states of energy between atoms and molecules became more predictable, complex, and accumulated. The nature of things meant a cascade of processes over millennia, with key steps and changes of state.

Protocells came from agglomerations of larger molecules. Multicellular life came from unicellular bacteria with boosted capacities when their attempts to absorb others met limits.[1] The history of life has come out of the relentless tension of "structure" versus "freedom."

Then there were events that changed the flow: human life emerged, long after an extraordinary group of reptiles was tested beyond its limits. Even though we are an unprecedented life-form, we are warned of being as dead as a dinosaur, as well as a dodo.

Limits are key to individual life. The cells in your body that exceed limits to growth are cancerous. Public health warnings help us to stay within limits of air quality, temperatures, drug and alcohol consumption, work hours and daily sleep. Limits and flourishing life go together.

We try to break limits to succeed but recognize care is needed. Take the tale of the tortoise and the hare, or the myth of Icarus who flew too high to the sun and melted the wax of his wings. Myths help us appreciate the deeper processes in human civilization that have brought us opportunities and crises, and they teach eloquently of limits. My intuition that people and place are bound to each other belongs to a deep preliterary history, along with the value of limits. Anthropologist David Abram eloquently described how the creation of the alphabet included a marked loss of connection to the land. Writing undermined a way of telling stories whereby landscapes and meaning were integrated.[2] I think writing made it easier to lose respect for the limits to life, and how those limits make life possible at all.

---

1. Margulis and Sagan, *Microcosmos*.
2. Abram, *Spell of the Sensuous*, 93–179.

The limits of our planet Earth also expose the limits of our abilities: we cannot agree on change, and many defend "business as usual." But the ability to write infinity does not mean life has no limits. These truths play out in your family, your life choices, your political preferences. Now is the time to act by truth.

Do you agree that apparent restrictions in your life have been vital, for good as well as bad? Olympic athletes set goals to push their limits by imposing rigorous training. Commitment of some sort is essential to live well. It is not "anything goes"; jobs have routines and relationships with some hierarchy and roles, and they give us identity and often meaning. Losing all this in the bliss of retirement can make it a death sentence unless new routines are created. From cradle to grave, we are embedded in social and economic contracts with clear limits and many fuzzy ones.

Our challenge is to take what we know as individuals to our systems and shared goals, especially where "free" is taken to be good, with no questions asked. It is to reject popularists who encourage lazy, knee-jerk reactions against taxes, bylaws, and policies to limit environmental harm. Our part is to explain to our neighbors why, and how, these changes are foundational: of a new order for a better future.

Christians dared to suggest that God set limits for Godself, in a charismatic teacher and healer. This God was self-limited and suffered. There would be no Jesus story if there had been no mortality, "even to death on a cross" (Phil 2:7–9). This limitation brought resurrected life to women and men, who in turn limited their lives, often to martyrdom. They grew in number and founded a new religion, changing human history. Christian faith embraces the limits of mortal life rather than trying to deny them. Any prosperity gospel of the faithful living as royalty, and problem free, has lost the plot.

Come back to the neutral ground of humanitarian concerns. What about your childhood? There was a time you were dependent on adults, and to be without them, terrifying. My childhood was relatively undramatic, but I have vivid memories of the first time I was taken to a summer fête with fun rides, sporting events, and music. There was so much to see. Adults were still very tall, and I lost my parents. The panic! Nothing horrible happened. My name was announced from the main tent, and my parents practically laughed at me when I cried in relief to see them. Take the limits of the family bubble away from a child, and all of life is shaken.

Grown up, we still find limits essential. Social media is a latest best thing but brings fraudulent profiles, harassment, content control, dependency, exaggeration, misinterpretation, and overinterpretation; all because it is a limited means of communication with the paradox of seemingly endless possibilities. The addictive power of social media, particularly with first-time users, meets the limits of the human body for sleep and daily functioning. Adjustments are needed, as long as we live online.

Music is a phenomenon that inspires a sense of freedom and well-being. Even the mournful fugue, or blues, is actually a statement of health and hope. Yet limits make music: the notes, sharp or flat, the pace, and the rhythm. The sound of music may be alive in the hills but only by contrasts, ricochets, and contexts.

Passion in sport is found through the paradox that the limits are crucial to competition. Take soccer and the goal-killing offside law. We can feel deep frustration in how it reduces play to the military precision of the defensive line. The "Offside!" call has ruined many fans' evenings. But the concept of "offside" came before soccer was formally born, in the nineteenth century. It was too easy to score, and agreement on limits meant greater satisfaction.

Food, our word for the daily digestion we must absorb to our bodies, must have limits of quantity, quality, and some balance. If we wish to improve food, we place more limits upon it: is it organic, fresh, and locally grown? Free of toxins, nuts, saturated fats? Prepared food is served artistically to inspire, reassure, relax, remind, seduce. Various limits bring life to our food, and life into us.

I have had fun thinking of the myriad forms and levels of human existence and how they show life is within limits; but do we notice?

The seventies "freedom" I experienced, of glorifying plastic, double garages, and weed-free lawns, is felt to be toxic today. Maybe this is still your neighborhood's culture, and Green is equated with less life rather than more? Social change is slow and intergenerational. Green initiatives provoke a "not in my backyard" attitude. Yet the truth of life within limits can persuade neighbors to welcome cycle paths, compost collection, or wind farms. Granted, they place limits on parking, where you put your kitchen waste, or your skyline. I have tried to describe how they make sense without reference to cash values because they correspond with life itself.

The movements to rewild and restore habitats are profound expressions of life within limits. If you have a garden with lawn, you could be part of wonderful restoration.[3] Larger-scale change will require your vote, your decision to limit your vote to parties that propose a more radical, "rooted in reality" agenda. But everything is linked. Changes in your garden will make conversations easier at those party conferences to fight out new policies. If progress can be made by the United Nations, as I witnessed at the COP15

---

3. Tallamy, *Nature's Best Hope*.

Biodiversity Conference, Montreal 2022, we are challenged to play our part too.

A new narrative for our shared life is emerging, but it is a battle, especially at national levels, where sufficient consensus is difficult. The reasons for this are more psychological than satanic! Sections of a population have differing worldviews, based on different stages of consciousness.[4] It is unfair just to blame politicians for being cautious when we live as a democracy with regular elections. How would someone possibly succeed if they did not give most of the population a familiar and reassuring message? Who in government would suggest planning for less "growth"? Not until popular mindsets shift.

Can you live by the truth of life within limits? Does it give you and the world you know more meaning and hope? The limits to both Green *and* Jesus can help.

The word "green" names the color we see in the chlorophyll of leaves, the reflected light that is *not* used in photosynthesis. For social change, Green is limited by blatant exploitation of our eco-illiteracy and the superficiality of it in most sectors. Yet Green grows in social power.

Jesus understood life within limits, with his parable that praises a dishonest manager who is uncovered by his billionaire boss: the manager has a chance to act before he loses his job. He calls his clients together and halves the debts of each. He knows he must rely on good human relationships rather than on cash (Luke 16:1–13).

This insight into relationships and money also highlights the depth of the problem we face. In the real world, countries as different as the USA and Haiti can never clear their debts. We live in systems that deny financial limits as much as the environmental ones. And that is no coincidence. The financial system sustains itself only by the interest charged upon debt, with the presumption that there is no limit to the purchasing power of hard cash. The impact of this is insidious. We were born into it, and the normality of it does not wish to be disturbed. To forgive debts as Jesus did could still get you crucified. Limitless money locks us into an ideology of growth. It is a lie that has been called out.[5]

I have referred to Norwegian teenager Greta Thunberg more than once, but did you see that dramatic scene when she said, "How dare you?" before the United Nations Climate Action Summit? It was an unforgettable moment of speaking truth to power:

---

4. See ch. 4, p. 98.
5. See ch. 4, "The Birth Pangs of Real Money," p. 101.

This is all wrong. I shouldn't be up here. I should be back in school on the other side of the ocean. Yet you all come to us young people for hope. How dare you!

You have stolen my dreams and my childhood with your empty words. And yet I'm one of the lucky ones. People are suffering. People are dying. Entire ecosystems are collapsing. We are in the beginning of a mass extinction, and all you can talk about is money and fairy tales of eternal economic growth. How dare you![6]

The emperor has no clothes. But the Hans Christian Andersen story has not yet come true, despite Greta. In the fable, false weavers sell their skills based on a lie that only the wise can see their work. Join me in outrage at anyone who smiles reassuringly about money and the status quo. Some cracks in the façade reached us the hard way with a housing bubble and the financial collapse of 2007–8. Meanwhile the Gretas of the world face suffocation.

There is a paradox that respect for the limits to life allows us to recreate it: whereas the folly of our time is to deny limits exist at all. We are warned to stay in limits, and to welcome the life they bring.

HOME TIME

Something new stirred within me, like baptism by the Spirit as flames and wind (Acts 1), when I realized Jesus was motivated by his understandings of the household. It was the power of the truth, a rebirth, as I appreciated how such a fundamental of life can reach into the convolutions of human systems, politics, and government. I scintillated with wonder and gratitude.

Home is prebiotic: before life. The home Earth found, with an orbit around the sun, nurtured life. Home runs through everything that lives but not like Russian dolls; it nurtures complexity and diversity, and our place in this is the articulation of everything Green.

Each spring, as the Canadian winter recedes, I see the power of coming home in dramatic migrations of geese who fly north and warblers who arrive for the budding of the trees. As summer comes, I remember how Montreal seduced me to migrate from the UK in 2001 with a luxurious "*joie de vivre*" of hot weather and festivals. Like birds and butterflies, I flew to make a new home, and I migrated back, often enough. What does home mean to you?

---

6. NPR Staff, "Transcript," paras. 3–4.

You may think of a building, but what about friendship with yourself, and your best friends who make you feel at home. The ability to reconnect, after years of absence, relies on knowing each others' deeper nature. In the moments that do not need words, the sense of being at home is priceless. This is home as "being known."

Home and being born belong together. It is a good sign if you take home for granted. A broken home is often the cause of adventurous stories of orphans, superheroes, villains, and saints who search for redemption. Home stands for the nurture each person needs, through parents, shelter, food, and pure air.

Canadians have begun to face the terrible impacts of the treatment of First Nations peoples, expressed through residential schools. Children were taken from their families and abused in a guise of betterment. It was cultural warfare through the home, just as some of my ancestors lost their homes in the Scottish-English history of the clearance of the Highlands: a vicious theft of house, land, language, and identity.

The yearning for home when things go wrong is often irresistible, but going home may not be good news. In myth, Odysseus has a ten-year journey to reach home. Even then, it ends badly, and it gave us the word "nostalgia." In 1983, I told a terrified young man to go home, when it meant his return to the civil war in Sri Lanka. His church had chosen him to share in a six-week cultural exchange visit. No one had planned for war, and I can picture him as he stood in the boarding zone of the airport, crying at his lack of choice. Forty years later, I realize our systems failed him; perhaps they still would. There are intrinsic conditions to home. He needed a new home, at least for a while.

What about turning people "on to the street" as their home? I was a volunteer worker at Streetwise Youth, a day center for young male sex workers in West London (1985–2002). This gave warmth, food, first aid, friendship, and advice. At the end of the day, a client chatted with me about anything they could think of, just to stay longer, because they had no home.

I picture my invalid grandma, widowed, staring out of the window to a nondescript street, on the poor side of Wolverhampton. She lived in a state-funded (council) home of two damp rooms, one fireplace, and a commode. Her youth was spent working the canals on a barge, poorly educated, and dependent on men: Victorian England, "the good old days"?

Jesus saw these sorts of things, and in faith, I know he cried, as I have done. The tears are authentic. It is wrong when home is lost, never known, or has harmful contradictions. Jesus' nativity stories suggest he did not have a normal sense of home. His Spirit-driven life was extreme towards his own family (Matt 12:48–50). Did home give him courage? Wilderness became his second home.

Can there be an antidote to the dread of global destruction, our planetary home? This huge existential question of our times can be answered from the lips of Jesus Green: "Today, salvation has come to this house" (Luke 19:9).

Salvation comes when we know our "house" has reached the ends of the earth.

WOMEN AND CHILDREN FIRST

A sense of looming catastrophe with climate change and species extinction fits the signs of a "second coming." Sometimes I wonder, if Jesus did return, would he be noticed? Has the "second coming" been missed, because folk were too busy looking at their little screens?

Every generation has its distractions; the sinking of the *Titanic* was due to a typical disregard for safety in that era of industrial achievement. Yet some details of this story point towards "salvation."

It is a true way to honor all those who have died, to remember the *Titanic* sank because it was going too fast to avoid an iceberg.[7]

The scene in James Cameron's film *Titanic* (1997), when Captain Smith accepts it is time for lifeboats, conveys how this was so mentally overwhelming. He thought of saving women and children first. This was no Hollywood gimmick: Cameron drew on the stories of survivors. But where did the phrase come from?

We have many details of the *Titanic* tragedy because over a thousand people survived to tell what happened. Captain Smith faced a terrible problem in abandoning ship. He knew there were too few lifeboats. What to do except to give some priorities to avoid chaos? He ordered, "Put the women and children in and lower away."[8] The two officers who heard him interpreted this differently, so some lifeboats were not full because all women and children waiting had boarded, while other boats then allowed men to fill each seat. Priorities made a difference: 74 percent of women passengers were saved, 52 percent of the children, and only 20 percent of the men.[9]

Shockingly, Captain Smith was not posthumously court-martialed, because his decisions reflected the procedures of his day. His order for

---

7. Five more lives were lost in 2023 when the explorer vessel the *Titan* imploded on its journey to the wreck of the *Titanic*.

8. Lord, *Night to Remember*, 43.

9. Anesi, "*Titanic* Disaster."

women and children was also conventional, based on a priority recalled in the shipwreck of HMS *Birkenhead* in 1852, thereafter known as the Birkenhead drill, thanks to a poem from Rudyard Kipling. The crew and soldiers followed orders to stand still, despite the hopeless situation of their shipwreck.

> To take your chance in the thick of a rush, with firing all about,
> Is nothing so bad when you've cover to 'and, an' leave an' likin' to shout;
> But to stand an' be still to the Birken'ead drill is a damn tough bullet to chew,
> An' they done it, the Jollies—'Er Majesty's Jollies—soldier an' sailor too!
> Their work was done when it 'adn't begun; they was younger nor me an' you;
> Their choice it was plain be'wee' drownin' in 'eaps an' bein' mopped by the screw,
> So they stood an' was still to the Birken'ead drill, soldier an' sailor too.[10]

Again, this was not the first time women and children were given priority. The earliest record comes from a "conflagration" of a packet boat, the *Poland*, May 16–18, 1840.[11]

Steaming from New York to Havre, almost mid-Atlantic, the *Poland* met a storm and was hit by lightening. Surely this was sudden death for all? Mr. J. H. Buckingham, a passenger from Boston, recalled the mental stress of the misfortune. The vessel withstood the shock, and further inspection confirmed it was basically sound. What joy, until someone smelled burning: a fire had caught in the cotton of the cargo hold. Attempts to fight it only gave more oxygen to the embers. Forced to retreat and wait, the deck began to feel warm. Evacuations were inevitable, but lifeboats for whom?

> On a suggestion that we might be obliged to take to the boats, it was immediately remarked by one of our French passengers and responded to by others—"Let us take care of the women and children first." I mention this as honorable to those who made it, and as showing that there was, even at that first moment of danger, a praiseworthy abandonment of self to the protection of others who are naturally more helpless.[12]

---

10. Kipling, "Soldier an' Sailor Too," stanza 5.
11. Howland, *Steamboat Disasters*, 334–51.
12. Howland, *Steamboat Disasters*, 341.

A longboat was put out with all women and children, and "as many men as felt proper, making thirty-five in all."[13] Sixty-four people remained on board. Could you have slept on the *Poland* when fire might break out at any moment? I would have given all my energy to scan the horizon. After two nights and three days of jeopardy, a larger ship, the *Clifton*, came to the rescue. A gale was blowing, but everyone was saved, even some cargo. Fire took over the *Poland* a few hours later. I am sure some would have noticed their salvation came on the third day, but ironically the women went through the worst: the longboat was cramped, exposed, and moved more in the water.

It was easy to give seven women on the *Poland* priority. But why? Perhaps chivalry or a primal sense to safeguard reproductive potential. It articulated the perception of what is still practiced in emergency teams worldwide. The vulnerable go first, and now it is not a matter of gender.

The sinkings of *Titanic*, *Birkenhead*, and *Poland* were exceptional. Most ships sink fast, and human behavior is far less generous. May 7, 1915, three years after the *Titanic* disaster, the passenger ship RMS *Lusitania* was eleven miles off the coast of Ireland when it was torpedoed by a German U-boat. She sank in eighteen minutes. The survivors were mostly ages sixteen to thirty. It was survival of the quickest! The lessons of the *Titanic* meant there were enough lifeboats, but only six were successfully launched. Calls for "women and children first" were overridden in the panic of survival instincts. For the women who escaped immediate drowning, they still faced sexual discrimination. The weight and restrictions to their clothing proved lethal. Proportionally many more women and children died than on the *Titanic*.[14]

"Women and children first" is clearer as a Jesus priority than as a practice in shipwrecks.

Jesus knew the Old Testament, where Sarah laughed at God, and God is more pragmatic and surprising than masculine brute strength.[15] Think of Joseph or Moses before Pharaohs, and David before Goliath, Samson and the Philistines, or Elijah and the prophets of Baal. Weakness becomes strength thanks to God, and the Messiah, the Son of David, has a lineage through women's friendships as much as men's (the book of Ruth).

Jesus takes a child and sets him in the middle of the disciples as an example of how to enter the realm of God (Matt 18:2–4). The loaves and fish Jesus takes to feed five thousand come from a young boy (John 6:9).

---

13. Howland, *Steamboat Disasters*, 336.
14. Parkinson, "Sisters in Fate"; Delap, "Shipwrecked."
15. Otwell, *And Sarah Laughed*, 54–55.

Jesus' famous "Let the little children come to me!" (Mark 10:14) follows a teaching against men discarding women by baseless divorce. Jesus rescues a woman from stoning (John 8:7). He was accessible to women; some shared their wealth with him (Luke 8:1–3), and he had a female following (Mark 15:41). A mother pleaded with Jesus for her daughter (Mark 7:26). Martha and Mary trust their brother Lazarus to him (John 11:1–3). A woman kisses Jesus' feet at the house of Simon the Pharisee (Luke 7:36–50), another is drawn to discuss personal life at a well (John 4:4–26). A story of the sisters Mary and Martha was remembered because he praised Martha, who listened to his teaching like a man would, rather than preparing food like Mary. These stories challenged the categorical thinking of the first- and second-century world.

Remember Jesus was homeless for a cause: he enjoyed hospitality, taught in homes and about households. All refer to the domains of female identity, where children were still second class.[16]

In contrast, the earliest Christian Scriptures, the letters sent by Paul for teaching what it means to be Christian, are typically categorical: "As in all the churches of the saints, women should be silent in the churches. For they are not permitted to speak, but should be subordinate, as the law also says. If there is anything they desire to know, let them ask their husbands at home. For it is shameful for a woman to speak in church" (1 Cor 14:33–34). And pointedly: "Wives, be subject to your husbands as you are to the Lord" (Eph 5:22). Social conformity applies to other relationships: "Slaves, obey your earthly masters in everything, not only while being watched and in order to please them, but wholeheartedly, fearing the Lord" (Eph 6:5). However, we cannot be sure all these words are Paul's, and there are other passages to argue Paul was an egalitarian. He works with, and applauds, women's contribution to the church, to the point of commissioning Phoebe, a deacon in the church (Rom 16).

Nevertheless, Jesus stands out as a disturbing teacher, who looked for new sorts of relationships between men, women, and children. Would he now question our relationship with nature?

In the 180 years since the *Poland* sank, the values expressed in shipwrecks have come home. Human rights have become enshrined in legal principles. Legal practice has enforced commonly held values of equality, dignity, and inclusion. It was an achievement in 1930 for the Canadian Parliament to pass a vote that "women are persons" and a logical step to accept nature as persons too. The process is well underway.

---

16. Malina, *New Testament World*, 46–57.

South African lawyer Cormac Cullinan has gathered together the story of the rights for nature global movement. He claims it will shape the twenty-first century as significantly as the human rights manifesto shaped the twentieth.[17]

Lawyer Christopher Stone wrote a landmark essay in 1972: "Should Trees Have Standing?—Toward Legal Rights for Natural Objects." This proposed legal personhood as a means of environmental protection. Montreal hosts the International Observatory on the Rights of Nature, founded in 2018.[18] Just as I found a collection of books and articles on gay rights helpful to my own liberation, those on the observatory website represent decades of human endeavors. They encourage a shift in perspective:

> Environmental justice theory appears, questioning the ethical duties of humans towards other species, and Nature's rights. Environmental justice moves away from the anthropocentrism, which leads development theory, and uses Ecocentrism instead. By the virtue of this new paradigm, man ceases to be the center of the world, and is thought to be just one of the many elements in Nature. From this viewpoint, all species and their lives are creditors of respect.[19]

Environmental disasters make the case for how we govern ourselves and the need for the movement to succeed. I find hope through this work. This paradigm includes the Universal Declaration of the Rights of Mother Earth on April 22, 2010. It tacitly acknowledges how the women's rights movement birthed this declaration. The International Joint Commission, a Canadian-US governmental organization for lake and river systems, offers a detailed Rights of Nature Timeline (1972–2019).[20]

In 2021, the Magpie River (*Muteshekau-shipu* in the Innu language), in northern Quebec, was granted legal personhood by the Innu Council of Ekuanitshit and the Minganie Regional County Municipality. Yenny Vega-Cárdenas, the president of the Observatory on the Rights of Nature, declared this a paradigm shift that went beyond protection of the river for future generations, as it changed the relationship between humans and the environment to something more complicated and intertwined.

Even though I have been committed to ecological concerns for many years I was unaware of these rights-based decisions. I knew Jesus put women and children first, and this alerted me to follow the process of inclusion and

17. Cullinan, *Wild Law*.
18. See https://observatoirenature.org/observatorio/en/home/.
19. Vega-Cárdenas and Parra, "Nature as a Subject," 128.
20. Slagle, "Rights of Nature FAQ," 3–5.

respect. In the violence to women and children that no longer passes as acceptable, there is also the recognition of violence to nature and a demand for more radical solutions.

This is where Jesus Green speaks, with passion, through the insights of the importance of home for all. Jesus knew the values of his Father's house outweighed all the money passing through it. He lives as Jesus Green with a call to enshrine "home for nature" in law, and the same Jesus said he came "not to undo the law but to fulfill it" (Matt 5:17).

Wild law is being manifested around the world, especially in New Zealand, but success is not guaranteed. In 2014 the town of Toledo, Ohio, experienced serious water pollution with toxic algae that made tap water undrinkable. Toledo declared Lake Erie a person in the Lake Erie Bill of Rights (LEBOR), February 26, 2019. This was struck down a year later by a federal court as "too broad."[21] The Magpie River is still not safe.[22] Despite the well-recorded impacts of hydroelectricity barrages, they are still considered as part of the Green transition in Quebec. Economic and ecological truths are dealt with by defining transition challenges in monetary terms.[23] First Nations are invited to be part of the process to increase hydroelectric capacity as never before, but in the language of dollars and cents. Laws need to be crafted to allow for predictable opposition and with reasonable caution for their success.

Nevertheless, this process of recognition of "nature as persons" is underway and will reach the places you know and love. It relies on principles that cannot be bounded by capital or national interests. It is biological, irresistible. On the streets in Montreal, sprayed graffiti declares "le future est antispéciste."

Life is found within the limits of the law, with the defense of the vulnerable, and legal progress can be the key to empower others. It certainly appears more reliable than politics and public opinion, against property developers who play a long game. In Quebec, I have seen interpretation of the law is as much a cultural phenomenon as other aspects of our common life. Hydro-Québec has tested public reaction with an action plan of many options. As before, the conservative-minded CAQ government hopes the remoteness of the river systems will mean weak opposition.

More case law is bound to shift the balance of power towards respect for nature.

---

21. Pallotta, "Federal Judge Strikes Down."
22. Dupuis, "Après la Romaine," 45:20–47:17.
23. Hamelin et al., "2035 Hydro-Québec Action Plan."

"Women and children first," and now nature, manifests a new form of common sense to change how things are done in practice.

HOME FOR ALL, NOT JUST FOR HUMANS

There is a fun parallel between stories of Jesus' ascension to heaven and contemporary space travel.

When the disciples were staring into the sky, as Jesus disappeared, they were met by men in white who asked why they were looking up: life would continue on earth (Acts 1).

When actor William Shatner, original *Star Trek*'s Captain Kirk, skirted Earth's outer atmosphere in the Blue Origin space shuttle, October 13, 2021, his world changed too:

> I love the mystery of the universe . . . all of that has thrilled me for years . . . but when I looked . . . I saw a cold, dark, black emptiness. It was unlike any blackness you can see or feel on Earth.
>
> I turned back toward the light of home. I could see the curvature of Earth, the beige of the desert, the white of the clouds and the blue of the sky. It was life. Nurturing, sustaining, life. Mother Earth. Gaia. Everything I had expected to see was wrong.
>
> I had thought that going into space would be the ultimate catharsis of that connection I had been looking for between all living things . . . I discovered that the beauty isn't out there, it's down here, with all of us. Leaving that behind made my connection to our tiny planet even more profound.
>
> The contrast between the vicious coldness of space and the warm nurturing of Earth below filled me with overwhelming sadness. Every day, we are confronted with the knowledge of further destruction of Earth at our hands: the extinction of animal species, of flora and fauna . . . things that took five billion years to evolve. . . . It filled me with dread. My trip to space was supposed to be a celebration; instead, it felt like a funeral.[24]

The first photo of an earthrise, taken by William Anders on Christmas Eve 1968, offered the world the proof of our common home. The French for a photo is *cliché*.

"One planet, one home, for all" is already a psychic reality, a cliché that cannot be undone. It is already a subconscious reality of the best possible future: now our minds and behavior are in a process of adjustment.

---

24. Shatner and Brandon, *Boldly Go*, 89–90.

We are disturbed by the truth of *home for all*. However eloquent the description, few of us celebrate our global home in our daily lives. It is far too abstract. Home for us is very particular, with many levels, and it is even more varied for nonhumans. But there is a common activity that can help us identify home, something far more down to earth: food.

Most of us enjoy home cooking. An animal's home is often where it eats: "where the deer and the antelope play."[25] Plants digest in place. A carnivore's hunting range can be their definition of home. Ancestral humanity survived through shelter and the warmth of the fire to provide cooked food and safer digestion. Meals conjure up memories, the smells and textures of childhood that can define our sense of home.

Many cultures draw out the relationships between bread and money, home and work: someone in the household has to find ways to "put bread on the table" (French: *mettre du pain sur la table*; Spanish: *pone el pan en la mesa*). Unless you are self-sufficient, food and money are inseparable. The Greek word for home or household, *oikos*, has given three key concerns: economics, ecology, and ecumenical.[26] The overlap between them highlights what is Green and questions any economics that denies homes rely on food, and food on the living world.

This is big-picture stuff. "Home" economics suddenly becomes more than cooking, and as vital as food. It throws the economic ideologies of the status quo into doubt, with broad strokes of first principles that can be integrated to the constitution of every nation.

Take the language of growth and the index GDP that is taken to measure it. Gross Domestic Product includes a hint of home with domesticity, but it was never intended to function as it now does. GDP was created as part of the US government program to overcome the Depression in 1937. It replaced Gross National Product (GNP), which included foreign activities. "Domestic" production was clearer and included citizens' activities overseas. An obvious problem for both GNP and GDP is how they fail to discriminate between productive and destructive products. Did you know the cost of habitat restoration after the oil pollution of the Deepwater Horizon disaster (2010) in the Gulf of Mexico is counted as a contribution to the GDP, likewise the war in Ukraine!? GDP also excludes most unpaid work. How can an increase in GDP be progress when it includes destructive impacts as positive, and is blind to volunteering?

A great summary to illustrate the true absurdity of GDP was put in personal terms:

25. "Home on the Range."
26. See ch. 4, p. [XREF].

So that the most "economically productive" citizen is a cancer patient who totals his car on his way to meet with his divorce lawyer.[27]

It is good to laugh, but we could cry for how this distorts our view of progress every day: a news broadcast on Radio Canada in 2021 featured how life in Lake Baikal, Siberia, was threatened by tourism, corruption, and inadequate infrastructures. It was followed by the daily economic feature, with joyful predictions of record GDP, post COVID-19. Our addiction to GDP parallels the alcoholic parent who creates a broken home.

The links between food, home, and the economy have been captured by award-winning economist Kate Raworth with *Doughnut Economics*. What's not to like about a doughnut? *Doughnut Economics* tempts us with a sweet contrast to limitless growth. It is a brilliant demonstration of the "without and the within of things" being practical. The two circles of a doughnut give the boundaries that lead to life within it. The outer ring is resource use and pollution, because of the physical limits of planetary life. The inner ring is the necessity to share these resources, so that social justice, health, and education are common rights to be ignored at our peril.

Doughnut Economics explores the dynamics of the *oikos* trinity. The reality of ecology is woven with an ecumenical perspective because Raworth has seen how this can work for the whole inhabited earth. She worked for years as the head economist for Oxfam, and she knows economics is drawn from reflections on the household:

> Ours is the era of the planetary household—and the art of household management is needed more than ever for our common home.[28]

Raworth proposes twenty-first-century *economics* places limits upon human activities. These will flourish between basic human needs as an inner limit, and environmental realities as an outer one.

Are you in the grip of economic ideology? Most households are in debt. Would your family react to "managing without growth" as something marginal or extreme? Peter A. Victor proposes just that. In *Managing without Growth*, Victor brings together wide-ranging and peer-reviewed research—walls of books—without falling into polemics. His review of the past two hundred years of ideas of growth shows this was never an exact

---

27. McKibben, *Deep Economy*, 28.
28. Raworth, *Doughnut Economics*, 49.

science, nor even a science: those who defend growth as unquestionable reveal themselves as suspect.[29]

Like Raworth, Victor takes on the dominance of GDP. It is measured through market prices, yet prices fail to match their theoretical principles. Each of the six classic requirements fails in practice. I was aware of one of these, the assumption that prices are not affected by externalities (e.g., noise pollution at airports), but prices also rely upon a homogeneity of products (branding impact is ignored), numerous participants (but often there are dominant sellers), freedom of entry and exit (companies can oppose the arrival of competitors), perfect information (are customers and sellers aware of all available products and their prices?), and finally, equal access to technology and resources (who thinks that is true in the real world?). The conclusion, that prices are inaccurate, permits a new view of the place of economics in political life. Victor observes policies can be price determining or price determined.

To put a dollar price on the value of a lake or a forest is full of problems. This is faux Green: it takes nature into the present flawed frameworks. Green economics redefines success as management to enhance life, rather than the ambiguous growth of productivity; this targets prosperity for all within the ecological limits of the planet.[30]

Victor argues growth is something to design away from because the predictors point to catastrophe: to be "slower by design, not disaster." Some growth is still needed in developing countries and this fits the *oikos principle* and *Doughnut Economics*. Basic standards to housing, diet, health care, education, and employment are goals to realize in every country. Yet richer countries have made this humanitarian growth more difficult. If the policy goals in richer countries moved away from growth, it would allow poorer countries to reach sustainable levels of income and welfare sooner. The growth goals for "rich" countries are not only increasingly incredible, but they also maintain terrible disparities between nations. Even then, success may feel empty: who needs a 4x4 to drive round town?

These necessities for a sustainable human future in the twenty-first century resonate with a story from Jesus of a rich man who had a bumper harvest: instead of sharing it, he decides to build bigger barns to keep the harvest for himself and have an easy life for many years. Only he dies suddenly, so all his building is worthless (Luke 12:15–21).

Both Raworth and Victor insist growth cannot be the definition of economic success. It seems to prevent sharing between humans when

---

29. Victor, *Managing without Growth*, 4–29.
30. See https://www.greeneconomycoalition.org.

sharing is vital. What we do with such wisdom of experts is key. Are you ready to change, to challenge people around you to let go of false certainties? Both authors refer to books I have from the 1980s, when I first took an interest in Green Economics, from Manfred Max Neef, Paul Ekins, Herman Daly, E. F. Schumacher, and George McRobie. Similar points, remade over thirty years later, show the inertia is within us, within family values, myths, and media messaging. What will it take? What drama or catastrophe? Can it be Greta at the United Nations? Floods and droughts in California? Wildfires across Europe? At each turn, toward more radical change, opposition is powerful. At the Climate Change COP28, Dubai 2023, a year of record temperatures, there was a quadrupling of participants who represented oil and gas companies.

Raworth compares our use of GDP with a cuckoo. Feeding GDP is the problem, and to lever GDP from the nest, we need an alternative, or more. We know biodiversity brings resilience to ecosystems; why not mimic this in measuring our economies? We can choose a democracy of economic language rather than a new emperor. A single indicator like GDP was suited to communicate globally across cultures, but today's logarithms are able to handle multiple values simultaneously.

A cuckoo is easier to eject than a global indicator linked to money. Alternatives wait for mainstream acceptance and some for decades; the Kingdom of Bhutan measures Gross National Happiness, The United Nations offers the Human Development Index (HDI), and a specialist nonprofit, the New Economics Foundation, has the Happy Planet Index. The Chinese government developed a Green GDP while a US initiative created the Genuine Progress Indicator. The Organization for Economic Cooperation and Development (OECD) proposes the Better Life Index.[31]

Each indicator has its history and certain weaknesses, but their intent is clear. The factors that leave us with GDP are tied up with a vision of progress that is over monetized and even claims coal is clean. What will give? The history of resistance to alternative indicators demonstrates the political realities of power, how we are governed, and the entanglement of the media.

Suppose that *"home for all, not just for humans"* is a natural expression of our survival instincts, a human expression of vital biofeedback, and part of the processes that have already brought billions of living things into existence. It is not going to disappear. Governance and the law can rely on it.

In chapter 4, "Deeper Green," I described how Teilhard de Chardin was prophetic.[32] His formations across geology, biology, philosophy, math-

---

31. Andester, "GDP Alternatives."
32. See ch. 4, p. 90–94.

ematics, and theology let him specialize as a geologist and paleontologist. He focused on genesis, the beginnings of species, what the fossil record can show us about the extraordinary history of life on Earth. When Teilhard became particularly concerned with human beginnings, he had the authority to write about human origins through the lens of life processes. He concluded the distinctive human identity is one of consciousness or knowledge.

Teilhard's main work, *Le phenomène humain* (1955), *The Phenomenon of Man* (1959), is a magnificent description of evolution that tells how humans came to exist as the "leading shoot" of life. He also made use of new vocabulary. Teilhard was the first to publish the term *noogenesis* (from the geochemist Vladimir Vernadsky): "The engendering and subsequent development of all stages of the mind" (Gk: *noos*);[33] and *noosphere*, for the global mind: "Outside and above the biosphere there is the noosphere." He concludes: "With hominisation . . . we have the beginning of a new age."[34]

Teilhard steers between optimism and despair on human nature without warning of the runaway impacts of human domination. Rather he writes with trust for the processes that have brought humanity to its global mind. As a phenomenon, humanity is extraordinary, but Teilhard had overlooked warning signs of extinctions, the dust bowl, and the true costs of industrialization. Two world wars are valid distractions, in his defense, and in one of his final publications he wrote of "the agony of our age: a world that is asphyxiating."[35]

The impacts of climate change, pollution, species loss, and ecosystem collapse bear upon the human phenomenon with catastrophic consequences. The new age risks being short. Our minds are challenged to accept different horizons for the noosphere; not the challenge to travel in space but something interior, of being human. We are conscious of limits that call us to do something, to avoid self-destruction. This cliché is a biological phenomenon through our minds: feedback that corresponds to innate reactions for survival. It brings ecumenical insights to economic systems: how to inhabit a common home.

"Home for all" describes and prophesizes, just as Teilhard pointed out the reality of the noosphere before the Internet. This can be understood across differences of education, intelligence, power, and livelihoods. It gives multiple qualities, without losing overall goals, and therefore offers a credible cohesion to "the Great Turning."[36] The next great step in planetary life is

---

33. Teilhard, *Phenomenon of Man*, 181.
34. Teilhard, *Phenomenon of Man*, 182.
35. Teilhard, *Activation of Energy*, 341–46.
36. See ch. 4, p. 107–108.

for our own minds and lives. We are challenged to accept different horizons in order to flourish. It is a process of reconciliation, to admit different interests and needs, to find success in compromises.

This is not a political step even though it has political demands. It is a biological phenomenon through our minds, and it is already a reality.

"Home for all" is the oxygen of human civilization: we have realized it is how we breathe. Notice the relief in your body, when you believe and trust this principle is the foundation for human civilizations.

Food can help us admit we are not passive participants in evolution. Homo sapiens is a worried top omnivore. We eat so much. Domestic livestock weigh many times the mass of the wild: 420 million tons for cows alone versus 22 million tons for wild land mammals.[37]

Consider the predator-prey relationship of wolves and deer. Levels of deer population can predict what will be the wolf population. Take away the wolves, and the deer population will rise, causing a loss of trees because more saplings are eaten. This leads to falling diversity of other species who are dependent on trees and falling soil quality with increased water runoff, as trees protect and renew soils. In the end the deer population itself can collapse due to the impacts of its own unchecked growth.

You may know the dramatic story of Pacific salmon and boreal forests in British Colombia, with bears who take thousands of salmon dinners from the river and fertilize the trees with their remains. The massive trees provide homes for hundreds more species. This way of describing the ecosystem, through who eats what and what happens to the remains, is the study of trophic flow (trophe = food). It demonstrates the interdependencies between species including humanity: home sweet home is a shared reality.

Look around. See what is real already. I notice "home for all, not just for humans" as a biological phenomenon with refuges for cows, pets, and wildlife; protected wilderness areas; removal of human-introduced species; plastic collection from land and oceans; coral research and restoration; rewilding by species and ecosystem restoration; residential garden trends to replace lawns with hospitable trees and shrubs; legal success to recognize nature as persons; guardians willing to die to protect the nature in their charge; "Give Nature a Home" as a slogan for bird protection in Scotland; international treaties to protect whales, oceans, biodiversity, and ozone round the Earth; municipal programs to plant indigenous trees, pollinator-friendly shrubs and flowers; buildings designed for trees and humans; reform of zoo policies towards species research and protection; national education programs on habitat and species protection; the 2021

---

37. Pennisi, "Who Rules the World," paras. 2, 7.

theme of the Christian Season of Creation: "A Home for All?"; blockbuster documentaries on struggles to save endangered species; critique of water management practices and improved sewer treatments; analysis of life cycles of products and industries, as well as organisms; battles to enforce the polluter pay principle (since 1972) with policies and organizations; national subsidies for agricultural reforms to favor biodiversity; better conditions for farmed animals and poultry; recognition and reductions in light pollution; climate change science; and controversy in its challenge to the status quo.

These are the manifestations of a shift of humanity away from species self-centeredness. They are the content behind the cliché, despite many other signs of individualism and egocentricity.

In the spirit of Teilhard de Chardin, and given his eloquent description of noogenesis and noosphere, I sum up these *beginnings* of "home for all" as *oikogenesis*. The global "home for all, not just for humans," is *oikogenesis*, for earth to manifest the *oikosphere*.[38]

Teilhard described evolutionary processes with radial and tangential energies and the human phenomenon, the arrival of mind, noogenesis, as predominantly a radial or expansive form: we dream of the stars.

With *oikogenesis* we are brought to earth, by the tangential energy of belonging. It has the necessary power to address our Achilles' heel, because for all our abilities we find it hard to change without violence. How does humanity achieve sufficient consensus across cultures, subcultures, and nationalities to achieve the different levels of change required for a sustainable world?

"Home" is self-evident yet universal. It is close to us, real; emotional as well as rational. Only such existential depth can guide rapid, multifaceted change. "Home for all" is translatable into many spheres of human activities. Again, this is not my being clever or persuasive. It is the boy who says, "The emperor is naked!" I name what is. "Home for all" grasps enough truth to accelerate efforts, at a time when acceleration is vital.

Mind is making home, just as we look for shelter when storm clouds appear. For many this is perceived as an "inconvenient truth" to their life plans, but "home for all" can work in the way "flat earthers" can see a ship through a telescope: the masts of a ship appear before the hull because it is blocked by the curvature of the water surface. No math is needed: you see it, and it cannot be unseen, just as the earthrise photo, or what William Shatner experienced, cannot be forgotten.

---

38. Wilson, *Half-Earth*.

"Home for all" allows the warnings of scientists to be re-expressed by politicians and planners, so our systems and agreements take account of the facts.

Most of us can grasp the sorts of things needed for the planet to be home for all. We have families who recycle, buy electric cars, and dig up the monoculture lawn. What do you do? Some of this is knowingly about being at home. Knowledge of the needs of plants, animals, insects, and birds grows thanks to media and eco-literacy. There is satisfaction to choose to live well. It is an energy that resonates with the sight of a clear-flowing spring or a shoal of fish seemingly without end. This is where the energy is radial, expansive. We will be profoundly happier, if . . .

There is a Dutch experience of homes designed to float on water. Amsterdam architects responded to anticipated rising water levels. The prototypes were good for humans but not for the aquatic life plunged into permanent shadow beneath. A redesign of homes, taken to a tropical context, was not just for humans: they act as artificial reefs to add to biodiversity of corals, fish, and invertebrates.[39]

Oikogenesis is not just an inconvenient truth; it is already expressed in actions. We are not at home: we cannot glory in disembodied mind, nor be happy when we can hear and see the screams of life being destroyed around us. Our earliest ancestors learned not to hold a hand in the fire, and the anguish of Greta Thunberg's "How dare you!" captured our unhappiness to the point of pain, so that we change.

We have begun to bring values to blind materialism, as a *new life process*. "Home for all" combines ecological truth with ecumenical know-how, for the many levels of human activities. It offers how to inhabit the earth without denying our human creativity and ambition. This makes *oikogenesis* the leading edge of human civilization through heart, mind, and stomach. It promises the happiness of "home" through biological truth and political plausibility. There is no time to lose.

AT HOME THROUGH THE CROSS

I admit that the power of the cross of Jesus has been abused and blunted. I ask you to give it a second chance. It is an absurd suggestion to be at home through the cross, whether the original one or the cross of Jesus Green. But Paul wrote of the folly of the cross, which encourages me to try (1 Cor 1:18).

---

39. Ward, "Is This Floating Eco-Pod," paras. 4–6; Hardingham-Gill, "Eco-Friendly Futuristic Floating Homes," para. 5.

I have attempted to set off a process within you, because Jesus Green relies on your input, your imagination. **If by chance you are skim-reading this book, stop here and go back. Give what I have written the chance to be integrated with your understandings *before* you reconsider the intensity of an ancient story.**

I realized the power of the cross by the spark I saw in people who handled Christian claims with humility and passion. Somehow, I knew they had connected with why Jesus died, and would die again.

Imagine a fresh crucifixion, awful as this is, to be open to the meaning Jesus offers in our time of need.

Here we find the cross of Jesus Green is prophetic and recreative. Religion has to let go of control to make room for the universal human yearnings for hope and courage. If this disturbs you, I plead that the cross is not a religious affair and never was. It was an awful punishment meant to eliminate threats to power. Now it is existential. Jesus died for our existence.

I was ten years old when my school took me to Boscobel House, Staffordshire, where Charles II hid in an oak tree during the English Civil War. Fighting also took place in Wales, Scotland, and Ireland, and more people died by proportion of the population than in the Great War of 1914–18.[40] In 1660, nine years after this epic escape from capture, Charles II succeeded his executed father, and the Boscobel oak became a pilgrimage site for royalists. Many wanted part of the tree as a souvenir. Branches and bark disappeared at a harmful rate, and despite efforts to protect the original tree it was gone by 1712.[41] A trade in oak-tree relics continued. Even in 2015 Christie's auctioneers offered an oak Queen Anne snuff box (c. 1710), with a silver-embossed drawing of Charles in the oak tree. It sold at over US$6000.[42]

I understand the patriotic sentiment for this oak: the wish to have a little piece of history; to touch its power. Even more for a relic of the cross. The story of Queen Helena, the mother of Emperor Constantine, is full of her good intentions to find the true cross. She visited Jerusalem around 290 years after Jesus died, which is not so much later than the recovery of a relic of the Boscobel oak today. Her party found three crosses and provoked a healing miracle to reveal which one bore Jesus. Too good to be true? Certainly! Perhaps the notice or titulus for Jesus that declared him "king of the Jews" could have been taken by Joseph of Arimathea. He had Pilate's approval to take responsibility for Jesus' body. But for the rest, the disciples were already afraid. Peter denied being with Jesus, and many left the city (to

40. Cartright, "Consequences," para. 6.
41. Cavendish, "Young Prince Hid," para. 6.
42. Dinerstein, "Queen Anne Silver."

go to Emmaus) or locked themselves in secure houses. Besides, why would anyone value those crosses? It was only after resurrection experiences, and then a spiritual transformation weeks later, that the group regained any sense of future. This future was the Living Lord, not blood-soaked locations of executions. Just as with Charles II, interest in relics of the cross would make sense only when it was safer and when there was a distance from the emotional trauma. No coincidence for the cross to be found after the conversion of the Roman emperor, a ban on crucifixions, and by his mother! Perhaps Queen Helena did not question or care about authenticity. She knew the value of her true cross lay in how it would inspire faithful imagination.

This is where the Gospels are helpful. They give us front-row crucifixion descriptions, by sound more than sight.

Jesus cries out, "My God, my God, why have you abandoned me?" (Matt 27:46). He draws on Ps 22. This psalm has a confident ending, which hints Jesus may have had hope as he died. I prefer the full-on abandonment because it is my experience, and the experience of most of us. If salvation is real, it must reach the unspeakably bad. The Gospels and Epistles also quote from Isa 53 to describe an afflicted servant of God, who is without relief in suffering for others.

But just like the creation of a relic, the meaning of the cross does not come from accurate historical details. It functions as a symbol of death and failure. Now as then,—and this failure was not simply an end but a beginning. It is flipped from defeat into victory. It responds to the human need for forgiveness, hope, and meaning.

The cross of Jesus Green emerges when we bring our needs to it. These are your personal needs, our needs as people who exploit and hurt each other, and the wounds to the earth as we deny the needs of other living beings.

Jesus died because he acted out a critique of the compromise Jewish leaders had made to operate the temple. The system had lost its integrity. Jesus refused to take power or plan for change. He simply exposed the corruption that power had brought. It is an exposure that every generation faces. Christianity succumbed to it when it became a religion of the powerful. The attraction of a relic may be a very personal expression of a love of power. While this is understandable, just as I take out travel insurance, it is not what Jesus offered.

> The Christian, who reads the Psalms with his attention on the cross, will find in the figure of the crucified the most radical and consistent critic of the religions of power. The life and death of

Jesus represent a refusal of privilege, a deliberate vulnerability, which subverts the triumphalist version of both Christianity and Judaism.[43]

Sadly, the God of power has been, for some, the God of power over others. Yet it was this God who died when Jesus died for us. It is something few preachers dare to press home. I know it makes most Christians uncomfortable, but I offer this because it has wider appeal in the twenty-first century. It is more like the crowds Jesus went to. It asks us to grow up, to let go of God as our spiritual comfort blanket whom we manipulate and so often bring down to self-concerns. I know Christian history has not kept to this, but it is there: "*My* God, why have you abandoned me?"

You accept that conditions of life would change if you went into extreme situations, like the bottom of the ocean or the top of a Himalayan mountain. Pressure means we cannot go to depths without a machine and technology. Everest ascents demand oxygen supplements or a certain carelessness for life. In all cases, we accept those who have gone through these experiences bring back valid truths. Yet they all came back to tell their tales.

Through his death, does the Spirit of Jesus speak to more than human failure? How and why would this be true? Suppose the Spirit of Jesus can place us in new relationships to everything, as his legacy resonates in the unprecedented expansion of human life on earth.

Two writers help prepare the ground for us. Both were profoundly changed by the atrocities of the twentieth century.

As well as being an outstanding theologian, Paul Tillich (1886–1965) was first a military chaplain in World War I and then a refugee of Nazi Germany. He was in the first group of university professors removed by Hitler's government in 1933. The emptiness of the eyes of those who went through trench warfare then became a fear Tillich expressed for his compatriots in wartime radio broadcasts. "He warned, that too much brutality and complicity with brutality would dehumanize the German people to such an extent that they would be looked on by the world as 'empty space and not living people. . . . In silent disregard there is the deepest rejection that a person can experience.'"[44]

In 1983, I visited Buchenwald concentration camp, southwest of Leipzig, and the German Methodist youth workers with me spoke of how they had questioned their parents who lived "over the hill, in the next valley." The pain of any adequate explanations, parent to child, threatened the same disregard. "Why had they done nothing?" These were churchgoing folk.

---

43. Shaw, *God in Our Hands*, 115.
44. Farrin, "Paul Tillich," para. 24.

Those deaths left a moral void. I understand how Tillich had the reputation as an apostle to the skeptics, because he created a dialog between religious life and probing reason. Now, he stands as a father for Jesus Green, to offer God without necessarily being religious. He goes beyond church, thanks to what is still found in the cross.

> The state of being grasped by the God beyond God is not a place where one can live, it is without the safety of words and concepts, it is without a name, a church, a cult, a theology. But it is moving in the depth of all of them. It is the power of being, in which they participate and of which they are fragmentary expression. . . . The courage to be is rooted in the God who appears when God has disappeared in the anxiety of doubt.[45]

I have every sympathy with an atheism that rejects the God of power. Instead, this appreciation of God beyond the God of power is present in the experience of the original crucifixion of Jesus: how his Spirit spoke of love, out of death.

Dr. Kosuke Koyama was born December 1929, in Tokyo and baptized as a Christian when he was fifteen in 1945, when American bombs rained down on Tokyo. Koyama had been struck by the courageous words of the presiding pastor, who told him that God called on him to love everybody, "even the Americans."[46] I first came across Dr. Koyama with his book *Water Buffalo Theology: A Culturalization of the Gospel for the Asian Context*. But it was his book *No Handle on the Cross* that marked me. Few theology books have drawings. He drew the hands present in three different traditions: Lenin's hand as a closed ideologically correct fist; the Buddha's webbed, open palm, of indiscriminate mercy; and Christ's nailed hand, neither open nor closed. Not open, in lacking definition or particularity, nor closed, in powerful confidence. The Christ hand bears the pain of the world. We are back to vulnerability and challenge. The shadow of the agonies of World War II and Koyama's response to love *even* the Americans reveal his was a "crucified mind." He contrasts the resourcefulness of Western Christianity (faith is like a lunch box) with the crucified mind that recognizes there is no handle on the cross. The cross is heavy and hard to carry. Godself has been like this from the beginning, when God asked, "Where are you?" to Adam and Eve: it shows God's respect for history (Gen 3:9). The question is repeated in varied ways through the primeval history of the first twelve chapters of Genesis. This is not God as the immune supreme being. Koyama

---

45. Tillich, *Courage to Be*, 190.
46. Martin, "Kosuke Koyama," para. 12.

suggests this insight comes from a crucified mind that confronts the deeper reality of God. The crucified mind is able to integrate truth and mercy:

> The crucified hands are the hands of ultimate love and respect for our history. They are the hands of divine invitation. The mind that contemplates the crucified hands, neither open nor closed, is the crucified mind. The crucified mind is perceptive about the varieties of forms of hands and their relationship to the crucified hands. . . . The crucified mind guides the positive energy of the "lunch-box" in the direction of the crucified Lord. The hands of Jesus are "Where-are-you? hands" of God.[47]

The death of God, as the supreme being, is a reading of the cross out of a respect for our history, which continues with Jesus Green. The "Where-are-you?" God will not let us ignore extinctions of species, just as Cain was unable to ignore the murder of his brother Abel.

The cross of Jesus Green is meaningful, with or without religion. It will take you from being a spectator to being a participant. To look on this cross risks your soul and sets your sails; for the sake of us all, from the blue whale to the humble sparrow. The church has sung this experience already:

> The Church of Christ in every age
> Beset by change but Spirit led,
> Must claim and test its heritage,
> And keep on rising from the dead.
>
> Across the world, across the street,
> The victims of injustice cry
> For shelter and for bread to eat,
> And never live until they die.[48]

For Jesus Green his cross is a symbol for our participation in oikogenesis. It evokes new life through our trust of the process to make "home for all."

Take the cross as a turning point, as the focus for our ultimate concerns. We can agree the dialogue between humanity and nature is of ultimate importance. It also has a paradox that marks out our attempts to reach for what is ultimate. The place of death can be the place of life.

Like the death of the tree to make the cross of Jesus' original execution, the death of the systems that preserve life on Earth are preparing a great crucifixion. The cutting away of balancing factors such as the path of the

---

47. Koyama, *No Handle on Cross*, 26.
48. Pratt Green, "Church of Christ," stanzas 1–2.

Gulf Stream, the rhythm of the monsoon, the pH of the ocean, the polar ice caps will also mean the death of civilization as we know it, and perhaps worse. This is a prediction of crucifixion as horrific and disobedient as the first.

One crucifixion was enough. This cross is an open symbol of crisis, of meaning over against despair, of judgment and reconciliation.

New meanings made from the cross have changed how artists portrayed it. In earliest pictures it was the victorious Christ, hardly affected by the nails, then the wounded suffering servant of the Middle Ages, and the astral Christ crucified of Salvador Dali in the twentieth, from the Spanish mystic St. John of the Cross, and popular as student posters.

The cross of Jesus Green is no different. But the events of the twentieth century with atomic bombs, and the rise of warfare and mass migration in the twenty-first, tell us God does not need to bring a second deluge: we know we can destroy ourselves.

There is a basic dynamic that cannot be lost in any authentic representation of Jesus' crucifixion. The hope. Somehow, what happened in the specific events of Jesus of Nazareth has truth for the whole of reality. The only reason we talk of the cross of Jesus is because of the new life that came from his dying. I have sensed the power to redecorate the cross with flowers on Easter day. It is deeper than the suburban love for happy endings. The church sings how the cross can move us to the promise of unbreakable, joyful love.

> O joy that seekest me through pain,
> I cannot close my heart to thee.
> I trace the rainbow through the rain,
> and feel the promise is not vain,
> that morn shall tearless be.
>
> O cross that liftest up my head,
> I dare not ask to fly from thee.
> I lay in dust, life's glory dead,
> and from the ground there blossoms red,
> life that shall endless be.[49]

The cross of Jesus Green reveals the animal, vegetal, and mineral in our salvation. This has been intuited over centuries. There is a stunning twelfth-century crucifixion mosaic in the half dome over the high altar of the Basilica of San Clemente al Laterano in Rome. It takes Jesus' statement "I am the true vine" (John 15). Jesus is on a simple cross, but around him

---

49. Matheson, "O Love," stanzas 3–4.

are doves roosting there, and the beloved disciple John and his mother Mary are either side. At the base of the cross is rich foliage that becomes branches in spirals reaching out across the entire ceiling. Birds, people, plants, and symbols find their place in the circling of the branches. There is a glory of gold broken up by the green of new life, and it is known as: *I Am the True Vine*, or *The Holy Tree of the Cross*. The image has been linked to later icons of Jesus, the True Vine, where the cross is almost assimilated yet assumed.

Jesus, the True Vine, **sixteenth century, Byzantine Museum, Athens**

These images show radical imagination. Christians dared to chase back their experience of the risen Jesus to the location of his death. They imagined God's purposes and picked up the vegetal life that Old Testament prophets took to describe human destinies. The stump of the tree gave life.

Some found it justified the divine rights of kings and queens to picture the Tree of Jesse, father of King David, in the lineage of Jesus (Isa 11:1).

This resting place is full of our questions and possibilities. I suggest a new abstraction of the crucifixion, for Jesus Green.

In the story *A Christmas Carol*, by Charles Dickens, a miser has his life turned upside down by the visit of three ghosts of Christmas, past, present, and future. It is a morality tale to question whether we live generously towards the world or not. Only the past, present, and future are not simply a morality tale. We know they are interrelated. I let the Jesus story replay to inspire a hopeful future.

The biosphere is in a crucified state because of human excesses. Greed and insecurities are at large, just as with Jesus of Nazareth. Jesus Green looks on the living world as his Father's house and rages that it is corrupted. He is found in all living things, not simply the human, so we look upon his crucifixion and find it judges us as executioners. Then it promises new life, via new boundaries, common agreements, and values. What we eat will be more important than money; our morality is informed by the truth of life being within limits, and life being shared like homes. This is the cross of reconciliation and recreation.

The cross of Jesus Green is the symbol of "home for all, not just for humans."

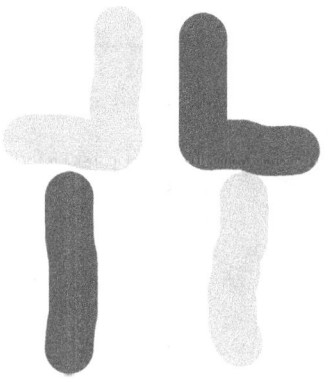

The Cross of Jesus Green
There are two, where there were once one.
Green is the tree, that became the cross:
all of nature that makes this planet Earth our home.
Red is the Savior and ourselves.[50]

50. See the color version on the back cover.

The power of the cross relies on both persons: nature and humanity. As separate single verticals, the dead tree kills the human. Brought together and entwined, there is a roof that speaks of a shared home. In the whole, I "see" this holy symbol as "hyper-stasis" or extreme stability. Such hyperbole addresses function: the paradox of a fixed symbol that demands change. We are forever in a relationship with nature, or we are homeless and dead. Nature does not hold grudges but waits for the opportunity to renew and restore. We saw this dynamic as nature moved into cities during the COVID pandemic. Through the reconciliation we make with nature, we find our true selves, at home, in our shared home. There is no going back, with the "hyper-stasis" of a new consciousness that weds us to the earth.

Hope is to recognize two, not one at the cross.

Picture the green and red in steady state.

Energy flows between them and beyond, the living world and the totality of humanity, our existence, all relationships. It is beyond my capacities to appreciate, but my imagination can dare to try. Daring changes me. The limits of the cross become limits that promise life, new consciousness, and joy for life. The cross of Jesus Green lets us face the possibility of human extinction with courage.

God save us from protecting our mediocrity, our denial that we are capable of murder, and the loss of ourselves.

I can imagine this symbol because it stands half in the past, with a prophet killed by powers of domination, and half in the present, as humanity and nature collide. There is a process to enter into, as with the original crucifixion. The Benedictine priest Sebastian Moore:

> *We* need the blood, not God.[51] . . .
> It is for one's acting out. It is designed to let him or her destroy one's wholeness and so to discover the love that is the indestructability of that wholeness. . . . One comes under that power only by shooting one's bolt, by exhausting one's power for an ultimate selfhood. . . . The identity that slowly flows into us to live with the crucified, is the life that, while it properly bears our name, is the climate in which others are finding their growing.[52]

To live with the crucified is to make homes for others. The cross of Jesus Green represents the holonic truth of existence: vegetal and animal are inseparable, out of mineral. We are part-wholes, and respect for this will let us flourish. Egos are part of the whole, and the human ego belongs to the earth. Ego is always in eco. This is *oikos*: we hang together. The selfhood

---

51. Moore, *Crucified Jesus*, 19; emphasis added.
52. Moore, *Crucified Jesus*, 52; emphasis added.

of humanity loses its destructive power and finds its true home. Glory be! The dais of the United Nations is to be open to the sky, and the seats of the delegates subject to the tides. Glory be!

Empty symbol or guide to life? We are invited to share in the suffering of species at risk, vulnerable human beings, and cultures. Nature and existence are big enough to take our despair and guilt: between nature and existence we can treasure life in new ways that make us truly happy. The eco-despair of millions who face needless loss and suffering becomes the hope of a new reconciliation.

When I participate in the cross of Jesus Green, I become aware of being alive at an extraordinary time. Despite the risks to our humanity, it gives courage to change. May I hold this cross to my heart and mind.

Jesus Green crucified says:

Happy are those who make home for all!

All-Age Epilogue
The Court of Nature

"To come to what we find the most compelling evidence, dear friends, the story of events since humans learned to make engines and devour the land and oceans." The crow spoke with sadness but clarity. Her eyes had seen enough death.

There was hesitation in the court. Who would speak? A wood pigeon landed clumsily on the dais. "I can tell you about the troubles of my family, but I have to speak of my cousins, the passenger pigeons, shot to extinction. They were so many. Flocks would take days to fly overhead. Forests would be bent over by their night roosting. Genocide. The killing did not stop."

Breathlessly, but so small, all strained to see her, a golden toad crawled up next: "I was never counted as important . . . global warming; we dried to death in Costa Rica!"

A colorful parakeet flew in, "People loved our feathers so much, we died for hats!"

"Your name?" asked the crow.

"Carolina parakeet—Audubon painted us!"

"This is not about color, I just happen to be orange," murmured the orangutan, "but you know I don't speak for myself or my kind alone. I speak for millions, the insects, undergrowth, epiphytes, fungi, great trees, monkeys, birds, we all are dying right now, right now despite the evidence on human screens across the globe, and what for? The oil in Nutella, in Pepsi, in pastry and fripperies! We are dying for fripperies. These humans are serial killers, eco-paths! Scum!"

The crow raised a wing, to gently remind the ape to keep to the oath of civility. By this time, the line-up to speak went beyond the courtroom. Ushers were bringing chairs. Most were weeping. Some looked angry.

The Tasmanian tiger said people noticed his passing, but today's extinctions are coming so fast they cannot even make the news; the Western black rhino, Japanese river otter, Cape Verde giant skink, the baiji dolphin in China: "And what about plants, the nobodies, like the St. Helena olive? Couldn't humans even protect an olive tree?"

A column of smelly spray fell upon the dais, turning heads to the Atlantic gray whale: "We were hunted to death as soon as the humans were able to get to open ocean in bigger boats, and the Atlantic has never been the same. You would have thought humans would change; they claim they can learn, but talk to the herring or the cod, mackerel or tuna. Have you seen the size of the latest factory ships? They are obscene."

So, it went on, the interplay between those gone for ever and those clinging to life, those forced to strange habitats and bizarre adaptations; the stories of predators being blamed when humans had disturbed the balance. Viruses were excluded for lack of time.

"Should the court continue?" cried the crow, still seeing the witness line without end. "Or is this talk through a full moon, and half more enough for a common mind?"

"Enough!" cried a voice, different than all others.

Opposite the crow in the dock were two humans. A Jew and a Christian. The Christian stood, white knuckled against the rail. "We said we are guilty two weeks ago, so why go on?"

"This is not so simple, nor your choice," replied the crow, "The story must be told, must be heard across centuries . . . and again, this is not just about you."

Only silence answered the crow's question. It was time.

"I ask both of the accused to stand and to reply clearly, loudly, and forever to our question: Those Homo sapiens in our court, how do you find your kind?"

"Guilty!" both men cried, empty with fatigue.

Coldly the crow continued, "Members of the court, what is the punishment merited by all that we have heard, and all that still could be said? Where is justice to be found, and how?"

The question came so soon after the verdict many were too stunned to react; others were more excited than ever, turning to their neighbor, chattering, shaking, jumping. A great horned owl flew high and round, calling for attention, acknowledged by the crow. "Your learned one, may we not use the wisdom of Homo sapiens to determine their own punishment, as they have confessed their guilt?"

"Go on."

"Take the story of the twelve tribes of Israel. Their father Jacob came to a critical point. As a rich man he returned home, to confront the brother he had cheated. On the eve of this meeting, he wrestled a man of God through the night to exhaustion. Jacob was injured at the hip yet asked for a blessing and received it. What sort of creature would wrestle with God and think he won? What is this Homo sapiens?

"For Christians, some say they are the ones who led the way in killing our world. Their teacher Jesus found his purpose in a deserted place, among us. He learned about healing and taught forgiveness. They say he was good, the best person ever. But what happened? He was killed horribly, like so many of us. What is this Homo sapiens?—They are arrogant, wicked, and dangerous!"

The owl's words left a heavy void. Each reflected on their own horrors. The crow broke the mood. "When we began this court, I described some options. Now tell us your decision.—Let the runners bring proposals to me."

Mice scurried at every level of the courtroom, reporting to monarch butterflies, who massed upon the crow, obscuring her completely, until revealed with both wings raised in solemn declaration.

"Not in the way of persons, but in the way of ourselves, from land, sea, and air, from the past, the present, and for the future; there is mercy from the court.

"You will return to your kind and tell them all you heard here, and you will change their minds, against the time that remains for us all.

"Christian, your story of Jesus comes to his execution. We have decided that the cross told you enough; it tells you that there would be mercy. Now you know it is true. You are pardoned, but tell your kind it was not just Jesus there. There was a tree whose life was taken, too, to make the cross. It was there all along. Now you recognize the truth. The tree is all of us, all of this court and more.

"We know the world of humans is changing. You say, 'Black Lives Matter,' now go beyond the human. Even the life of a sparrow matters, like Jesus said.

"Go and change your minds, your ways, and naturalize your systems, to practice mercy as you have received it: before it is too late. Take the risks you must!"

And all the court cried, sang, croaked, and trembled with amen.

> *Do not get lost in a sea of despair. Be hopeful, be optimistic. Our struggle is not the struggle of a day, a week, a month, or a year, it is the struggle of a lifetime. Never, ever be afraid to make some noise and get in good trouble, necessary trouble.*
> Rep. John Lewis, black rights activist (1940–2020)
> Tweeted June 27, 2018

Bibliography

Abram, David. *Becoming Animal: An Earthly Cosmology*. New York: Vintage, 2011.
———. *The Spell of the Sensuous: Perception and Language in a More-Than-Human World*. 2nd ed. New York: Vintage, 2017.
Adams, Douglas. "The answer to this is very simple." alt.fan.douglas-adams, Nov. 3, 1993. https://web.archive.org/web/20080201064612/https://groups.google.com/group/alt.fan.douglas-adams/msg/d1064f7b27808692?dmode=source.
———. *The Hitchhiker's Guide to the Galaxy*. London: Pan, 1979.
Alison, James. *Faith beyond Resentment: Fragments Catholic and Gay*. New York: Herder & Herder, 2001.
Anderson, Bernhard W. *The Living World of the Old Testament*. 4th ed. Harlow, UK: Longman, 1988.
Andester, Nikita. "GDP Alternatives: 7 Alternative Ways to Measure a Country's Wealth." Ethical Network, n.d. https://ethical.net/politics/gdp-alternatives-7-ways-to-measure-countries-wealth/.
Anesi, Chuck. "*Titanic* Disaster: Official Casualty Figures." Anesi, 1997. https://www.anesi.com/titanic.htm.
Annaud, Jean-Jacques, dir. *The Name of the Rose*. [Italy:] Columbia, 1986.
Apted, Michael, dir. *Amazing Grace*. N.p.: Momentum, 2007.
Arcand, Denys, dir. *Jesus of Montreal*. Toronto: Cineplex Odeon, 1989.
Armstrong, Karen. *Sacred Nature: Restoring Our Ancient Bond with the Natural World*. New York: Knopf, 2022.
Audubon, John James. *The Birds of America: From Drawings Made in the United States and Their Territories*. 8 vols. New York: Chevalier, 1840–42.
Balabanski, Vicky. "Critiquing Anthropocentric Cosmology." In *Exploring Ecological Hermeneutics*, edited by Norman Hegel and Peter Trudinger, 151–59. Atlanta: Society of Biblical Literature, 2008.
Barrett, C. K. *The Gospel According to St John: An Introduction with Commentary and Notes on the Greek Text*. 2nd ed. London: SPCK, 1978.
Bauckham, Richard. *The Bible and Ecology: Rediscovering the Community of Creation*. London: DLT, 2010.
BBC News. *Our World*. BBC News, July 5, 2021. https://archive.org/details/BBCNEWS_20210705_003000_Our_World/start/720/end/780.
Berry, Thomas. *The Dream of the Earth*. Berkeley, CA: Counterpoint, 1988.
Besson, Luc, dir. *The Big Blue*. Neuilly-sur-Seine, Fr.: Gaumont, 1988.
Binet-Sanglé, Charles. *La folie de Jésus*. 3 vols. Paris: Maloine, 1900–1912.

Böll, Heinrich. *Meat Atlas*. Heinrich Böll Foundation, 2021. https://www.boell.de/en/tags/meat-atlas.

Boswell, James. *Christianity, Social Tolerance, and Homosexuality: Gay People in Western Europe from the Beginning of the Christian Era to the Fourteenth Century.* Chicago: University of Chicago Press, 1980.

Bradley, Ian. *God Is Green: Christianity and the Environment.* 2nd ed. London: DLT, 2020.

Bress, Eric, and Makye J. Gruber, dirs. *The Butterfly Effect.* Burbank, CA: New Line Cinema, 2004.

Brewster, David. *Memoirs of the Life, Writings, and Discoveries of Sir Isaac Newton.* 2 vols. Cambridge: Cambridge University Press, 1855.

Brown, Peter G., and Geoffrey Carver. *Right Relationship: Building a Whole Earth Economy.* San Francisco: Berrett-Koehler, 2009.

Buber, Martin. *I and Thou.* Centennial ed. New York: Scribner's, 2023.

Butterfield, Roger. "Henry Ford, the Wayside Inn, and the Problem of 'History Is Bunk.'" Bunk History, June 1, 1965. https://www.bunkhistory.org/resources/henry-ford-the-wayside-inn-and-the-problem-of-history-is-bunk.

Cameron, James, dir. *Avatar.* Los Angeles: 20th Century, 2009.

———. *Titanic.* Los Angeles: Paramount, 1997.

Canadian Nuclear Safety Commission. "Low- and Intermediate-Level Radioactive Waste." Nuclear Safety, last modified May 4, 2021. http://www.nuclearsafety.gc.ca/eng/waste/low-and-intermediate-waste/index.cfm#intermediate-level.htm.

Capra, Fritjof. *The Tao of Physics: An Exploration of the Parallels between Modern Physics and Eastern Mysticism.* London: Wildwood, 1975.

Capra, Fritjof, and Pier Luigi Luisi. *The Systems View of Life: A Unifying Vision.* Cambridge: Cambridge University Press, 2014.

Carroll, Sean B. "The Ecologist Who Threw Starfish." Nautilus, Mar. 7, 2016. https://nautil.us/the-ecologist-who-threw-starfish-235831.

Carson, Rachel. *Silent Spring.* Boston: Houghton Mifflin, 1962.

Cartright, Mark. "Consequences of the English Civil Wars." *World History Encyclopedia*, Feb. 9, 2022. https://www.worldhistory.org/article/1944/consequences-of-the-english-civil-wars.htm.

Cavendish, Richard. "The Young Prince Hid from Roundhead Soldiers on September 6th, 1651." *History Today* 51 (2001). https://www.historytoday.com/archive/charles-ii-hides-boscobel-oak.

Conradie, Ernst M., et al, eds. *Christian Faith and the Earth: Current Paths and Emerging Horizons in Ecotheology.* London: Bloomsbury T&T Clark, 2014.

Cowell, Alan. "After 350 Years, Vatican Says Galileo Was Right: It Moves." *New York Times*, Oct. 31, 1992. https://www.nytimes.com/1992/10/31/world/after-350-years-vatican-says-galileo-was-right-it-moves.html.

Cullinan, Cormac. *Wild Law: A Manifesto for Earth Justice.* 2nd ed. White River Junction, VT: Chelsea Green, 2011.

Curie, Marie. *Pierre Curie.* Translated by Charlotte Kellogg and Vernon L. Kellogg. New York: Macmillan, 1923.

Daly, Herman E., and John B. Cobb Jr. *For the Common Good: Redirecting the Economy toward Community, the Environment, and a Sustainable Future.* Boston: Beacon, 1989.

Danzin, André, and Jacques Masurel. *Teilhard de Chardin, visionnaire du monde nouveau*. Monaco: Rocher, 2005.

Darwin, Charles R. *The Formation of Vegetable Mould through the Action of Worms*. London: Murray, 1881.

Delap, Lucy. "Shipwrecked: Women and Children First?" University of Cambridge, Jan. 20, 2012. https://www.cam.ac.uk/research/discussion/shipwrecked-women-and-children-first.

Dengler, Roni. "Neonicotinoid Pesticides Are Slowly Killing Bees." PBS, June 29, 2017. https://www.pbs.org/newshour/science/neonicotinoid-pesticides-slowly-killing-bees.

Devall, Bill, and George Sessions. *Deep Ecology: Living as if Nature Mattered*. Layton, UT: Smith, 1985.

Dickens, Charles. *A Christmas Carol: In Prose; Being a Ghost Story of Christmas*. London: Chapman & Hall, 1843.

———. *Oliver Twist*. London: Dent and Sons, 1907.

Dinerstein, Nick. "A Queen Anne Silver and Oak Snuff Box of Historical Interest, circa 1710." Christies, May 20, 2015. https://www.christies.com/en/lot/lot-5896633.

Dinkler, Michal Beth. "The Bible and Women? We Need to Talk." *Reflections*, 2019. https://reflections.yale.edu/article/resistance-and-blessing-women-ministry-and-yds/bible-and-women-we-need-talk.

Discus. "The Origins of the Pigeon Blood Discus Strain." Discus, n.d. https://www.discus.com/science/the-origins-of-the-pigeon-blood-discus-strain.

Drury, John. *The Parables in the Gospels: History and Allegory*. London: SPCK, 1985.

Duignan, Brian. "Democritus: Greek Philosopher." *Britannica*, last updated Feb. 27, 2024. https://www.britannica.com/biography/Democritus.

Dunn, James D. G. *Christology in the Making: A New Testament Enquiry into the Origins of the Doctrine of the Incarnation*. 2nd ed. Grand Rapids: Eerdmans, 1989.

———. *Did the First Christians Worship Jesus? The New Testament Evidence*. London: SPCK, 2010.

———. *The Theology of Paul the Apostle*. Edinburgh: T&T Clark, 1998.

Dupuis, Roy, presenter. *Doc humanité*. Season 5, "Après la Romaine." Aired Sept. 23, 2023, on Radio-Canada. https://ici.radio-canada.ca/tele/doc-humanite/site/episodes/807086/roy-dupuis-riviere-magpie-hydroelectricite.

Eco, Umberto. *The Name of the Rose*. Translated by William Weaver. New York: Harcourt, 1983.

Ehrman, Bart D. *How Jesus Became God: The Exaltation of a Jewish Preacher from Galilee*. New York: HarperOne, 2014.

Eiesland, Nancy L. *The Disabled God: Toward a Liberation Theology of Disability*. Nashville: Abingdon, 1994.

Einstein, Albert. *The Ultimate Quotable Einstein*. Collected and edited by Alice Calaprice. Princeton, NJ: Princeton University Press, 2010.

Eisenstein, Charles. *Sacred Economics: Money, Gift & Society in the Age of Transition*. Berkeley, CA: Evolver, 2011.

Farrin, Cassandra J. "Paul Tillich, 50 Years Later." Westar Institute, Nov. 5, 2015. https://www.westarinstitute.org/blog/paul-tillich-50-years-later.

Fiorenza, Elisabeth Schüssler. *In Memory of Her: A Feminist Reconstruction of Christian Origins*. London: SCM, 1983.

Fontana, Tom. *Oz*. Aired July 12, 1997–Feb. 23, 2003, on HBO.

Fowler, James W. *Stages of Faith: The Psychology of Human Development and the Quest for Meaning*. New York: HarperOne, 1981.

Francis, Pope. "*Laudato Si'*: On Care for Our Common Home." Vatican, May 24, 2015. https://www.vatican.va/content/francesco/en/encyclicals/documents/papa-francesco_20150524_enciclica-laudato-si.html.

Fries, Theodor M. *Linnaeus*. Edited and translated by Benjamin D. Jackson. Cambridge Library Collection. Repr., Cambridge: Cambridge University Press, 2011.

Fromm, Eric. *The Heart of Man: Its Genius for Good and Evil*. New York: Harper and Row, 1964.

Fuller, Errol. *The Passenger Pigeon*. Princeton, NJ: Princeton University Press, 2014.

Fuller, Reginald H., and Pheme Perkins. *Who Is This Christ? Gospel Christology and Contemporary Faith*. Philadelphia: Fortress, 1983.

Funk, Hoover, et al. *The Five Gospels: The Search for the Authentic Words of Jesus*. New York: Harper Collins, 1997.

George, Henry. *Progress and Poverty: An Inquiry into the Cause of Industrial Depressions and of Increase of Want with Increase of Wealth: The Remedy*. New York: Appleton, 1879.

Gesell, Silvio. *The Natural Economic Order*. Translated by Philip Pye. Berlin: NEO, 1906. https://www.community-exchange.org/docs/Gesell/en/neo.

Gibson, Mel, dir. *The Passion of the Christ*. Los Angeles: New Market, 2004.

Glover, Raymond F. *The Hymnal 1982 Companion*. New York: Church, 1990.

Goldberg, G. J. "Evaluating the Josephus-Jesus Paraphrase Model." A Flavius Josephus Blog, June 19, 2022. https://josephusblog.org.

Goleman, Daniel, et al. *Eco-Literate: How Educators Are Cultivating Emotional, Social, and Ecological Intelligence*. San Francisco: Jossey-Bass, 2012.

Goss, Robert. *Jesus Acted Up: A Gay Manifesto*. New York: Harper Collins, 1994.

Grof, Stanislav. *The Holotropic Mind: The Three Levels of Human Consciousness and How They Shape Our Lives*. New York: Harper San Francisco, 1990.

Guevara-Mann, Pedro, and Aaron Flanzraich. "In the Beginning: Signs and Symbols." YouTube, Nov. 25, 2016. https://www.youtube.com/watch?v=JaWYrpkd3pE.

Gutiérrez, Gustavo. A *Theology of Liberation: History, Politics and Salvation*. New York: Orbis, 1988.

Habel, Norman C. *An Inconvenient Text: Is a Green Reading of the Bible Possible?* Adelaide: ATF, 2009.

Habel, Norman C., and Peter Trudinger, eds. *Exploring Ecological Hermeneutics*. Symposium. Atlanta: Society of Biblical Literature, 2008.

Hamelin, Paule, et al. "2035 Hydro-Québec Action Plan: Québec Is Ramping Up Its Energy Transition Efforts." Gowling WLG, Nov. 29, 2023. https://gowlingwlg.com/en/insights-resources/articles/2023/2035-hydro-quebec-action-plan-energy-efforts.

Hardingham-Gill, Tamara. "These Eco-Friendly Futuristic Floating Homes Are Currently under Construction." Waterstudio.NL, Aug. 16, 2022. https://www.waterstudio.nl/these-eco-friendly-futuristic-floating-homes-are-currently-under-construction.

Harvey, Andrew, and Anne Baring. *The Mystic Vision: Daily Encounters with the Divine*. Alresford, UK: Godsfield, 1995.

Heermann, Johann. "Ah, Holy Jesus, How Hast Thou Offended." Translated by Robert S. Bridges. In *Hymns and Psalms: A Methodist and Ecumenical Hymn Book*, #164. London: MHP, 1983.

Henry, Patrick. *New Directions in New Testament Study*. London: SCM, 1980.

Hill, Jim, and Rand Cheadle. *The Bible Tells Me So: Uses and Abuses of Holy Scripture*. New York: Anchor, 1996.

Hiroshima for Global Peace. "Meeting Trees: Sending Trees That Survived Atomic Bombing, from Hiroshima to the World." Hiroshima for Global Peace, n.d. https://hiroshimaforpeace.com/en/sending-trees-that-survived-atomic-bombing-from-hiroshima-to-the-world.

Hirst, Michael. *The Tudors*. Aired 2007–10, on BBC Two.

"Home on the Range." Wikipedia, last edited Feb. 20, 2024. https://en.wikipedia.org/wiki/Home_on_the_Range.

Horrell, David G., et al., eds. *Ecological Hermeneutics: Biblical, Historical and Theological Perspectives*. New York: T&T Clark, 2010.

———, eds. *Greening Paul: Rereading the Apostle in a Time of Ecological Crisis*. Waco, TX: Baylor University Press, 2010.

Horsley, Richard, A. *Jesus and Empire: The Kingdom of God and the New World Disorder*. Minneapolis: Fortress, 2003.

Howland, Southworth Allen. *Steamboat Disasters and Railroad Accidents in the United States*. Worcester, MA: Lazell, 1840.

Jeremias, Joachim. *Jerusalem in the Time of Jesus: An Investigation into Economic and Social Conditions during the New Testament Period*. Philadelphia: Fortress, 1969.

———. *The Parables of Jesus*. New York: Scribner's, 1972.

Johnson, Elizabeth A. *Ask the Beasts: Darwin and the God of Love*. London: Bloomsbury, 2014.

Jones, Terry, dir. *Monty Python's Life of Brian*. London: Cinema International, 1979.

Josephus, Flavius. *Josephus: The Complete Works*. Translated by William Whiston. Nashville: Thomas Nelson, 1998.

Kaneda, Toshiko, and Carl Haub. "How Many People Have Ever Lived on Earth?" PRB, Nov. 15, 2022. https://www.prb.org/articles/how-many-people-have-ever-lived-on-earth.

Kearney, Richard. *Anatheism: Returning to God after God*. New York: Columbia University Press, 2010.

Kegan, Robert. "What 'Form' Transforms? A Constructive-Developmental Approach to Transformative Learning." In *Learning as Transformation: Critical Perspectives on a Theory in Progress*, edited by Jack Mezirow, 35–69. San Francisco: Jossey-Bass, 2000.

Kennedy, Charles W. "A Dream of the Cross." In *Earliest English Christian Poetry: Translated into Alliterative Verse, with Critical Commentary*, 93–97. London: Hollis and Carter, 1952.

Kimmerer, Robin Wall. *Braiding Sweetgrass: Indigenous Wisdom, Scientific Knowledge, and the Teachings of Plants*. Minneapolis: Milkweed, 2013.

Kipling, Rudyard. "Soldier an' Sailor Too." *Pearson's Magazine* 1 (Apr. 1896) 386.

Klein, Naomi. *This Changes Everything: Capitalism vs. the Climate*. London: Lane, 2014.

Kormandy, Edward J. "Ecology/Economy of Nature—Synonyms?" Review of *Nature's Economy*, by Donald Wooster. *Ecology* 59 (1978) 1292–94.

Korton, David C. *The Great Turning: From Empire to Earth Community*. Oakland, CA: Berrett-Koehler, 2006.

Koyama, Kosuke. *No Handle on the Cross: An Asian Meditation on the Crucified Mind*. Repr., Eugene, OR: Wipf and Stock, 2011.

László, Ervin. "Integral Consciousness in a Self-Actualizing Cosmos." YouTube, Feb. 10, 2015. From introductory lecture of First Integral European Conference, 2014. https://www.youtube.com/watch?v=wJ9TJAISrLw.

Law, William. *A Serious Call to a Devout and Holy Life*. London: Collins, 1965. First published 1728.

Lederman, Leon M., with Dick Teresi. *The God Particle: If the Universe Is the Answer, What Is the Question?* New York: First Mariner, 2006.

Leopold, Aldo. *A Sand County Almanac*. New York: Ballantine, 1966. First published 1949.

Loorz, Victoria. *Church of the Wild: How Nature Invites Us into the Sacred*. Minneapolis: Broadleaf, 2021.

Lord, Walter. *A Night to Remember*. New York: Bantam, 1997.

MacCulloch, Diarmaid. *Christianity: The First Three Thousand Years*. New York: Penguin, 2009.

Macy, Joanna. "Foreword." In *Stories of The Great Turning*, edited by Peter Reason and Melanie Newman, 5–8. London: Kingsley, 2017.

———. "Joanna Macy on the Great Turning." YouTube, 2005. https://www.youtube.com/watch?v=LwlXTAT8rLk&t=265s.

Macy, Joanna, and Sally Young Brown. *Coming Back to Life: Practices to Reconnect Our Lives, Our World*. 9th ed. Gabriola Island, BC: NSP, 2011.

Maier, Paul L. "Josephus on Jesus." Issues, Etc., n.d. From *Josephus: The Essential Works*. https://www.issuesetcarchive.org/articles/bissar24.htm.

Makower, Joel. "The Green Consumer, 1990–2010." GreenBiz, Mar. 29, 2010. https://www.greenbiz.com/article/green-consumer-1990-2010#.

Makower, Joel, et al. *The Green Consumer: You Can Buy Products That Don't Cost the Earth*. London: Penguin, 1990.

Malina, Bruce J. *The New Testament World: Insights from Cultural Anthropology*. 3rd ed. Louisville: Westminster John Knox, 2001.

Malthus, Robert Thomas. *An Essay on the Principle of Population*. London: Johnson, 1798.

Margulis, Lynn, and Dorion Sagan. *Microcosmos: Four Billion Years of Microbial Evolution*. New York: Summit, 1986.

Márquez, Gabriel García. *Love in the Time of Cholera*. Translated by Edith Grossman. New York: Vintage, 1988.

Martin, Douglas. "Kosuke Koyama, 79, Ecumenical Theologian, Dies." *New York Times*, Apr. 1, 2009. https://www.nytimes.com/2009/04/01/world/asia/01koyama.html.

Matheson, George. "O Love That Wilt Not Let Me Go." In *Hymns and Psalms: A Methodist and Ecumenical Hymn Book*, #685. London: MHP, 1983.

Matulka, Rebecca. "The History of the Electric Car." Energy.gov, Sept. 15, 2014. https://www.energy.gov/articles/history-electric-car.

McCarthy, Michael. *The Moth Snowstorm: Nature and Joy*. London: Murray, 2015.

McCormick, John. *The Global Environmental Movement*. Hoboken, NJ: Wiley, 1995.

McFague, Sallie. *The Body of God: An Ecological Theology*. Minneapolis: Fortress, 1993.

McKibben, Bill. *Deep Economy: The Wealth of Communities and the Durable Future.* New York: St. Martin's Griffin, 2007.

McKintosh, Steve. *Integral Consciousness and the Future of Evolution: How the Integral Worldview Is Transforming Politics, Culture and Spirituality.* Saint Paul: Paragon, 2007.

Meadows, Donella H., et al. *The Limits to Growth: A Report for the Club of Rome's Project on the Predicament of Mankind.* New York: Universe, 1972.

Meggitt, Justin J. "The Madness of King Jesus: Why Was Jesus Put to Death, but His Followers Not?" *Journal for the Study of the New Testament* 29 (2007) 379–413.

Mendes, Sam, dir. *American Beauty.* Universal City, CA: DreamWorks, 1999.

Mill, John Stuart. *A System of Logic, Ratiocinative and Inductive: Being a Connected View of the Principles of Evidence, and the Methods of Scientific Investigation.* London: Cambridge University Press, 1846.

Miller, William. "Old Man's Advice to Youth: 'Never Lose a Holy Curiosity.'" *LIFE Magazine* (May 2, 1955) 64.

Moore, Sebastien, OSB. *The Crucified Jesus Is No Stranger.* Rev. ed. New York: Paulist, 1977.

Mosala, Itumeleng J., and Buti Tlhagale, eds. *The Unquestionable Right to Be Free: Black Theology from South Africa.* Maryknoll, NY: Orbis, 1986.

Mueller, John D. *Redeeming Economics: Rediscovering the Missing Element.* Wilmington, DE: ISI, 2010.

Myers, Ched. *The Biblical Vision of Sabbath Economics.* Washington, DC: Tell the Word, 2001.

———. *Binding the Strong Man: A Political Reading of Mark's Story of Jesus.* New York: Orbis, 2000.

Neusner, Jacob. *Jews and Christians: The Myth of a Common Tradition.* London: SCM, 1991.

Nevett, Joshua. "How Green Politics Are Changing Europe." BBC, Oct. 21, 2021. https://www.bbc.com/news/world-europe-58910712.

Noonan, David. "The 25% Revolution—How Big Does a Minority Have to Be to Reshape Society?" *Scientific American*, June 8, 2018. https://www.scientificamerican.com/article/the-25-revolution-how-big-does-a-minority-have-to-be-to-reshape-society/.

NPR Staff. "Transcript: Greta Thunberg's Speech at The U.N. Climate Action Summit." NPR, Sept. 23, 2019. https://www.npr.org/2019/09/23/763452863/transcript-greta-thunbergs-speech-at-the-u-n-climate-action-summit.

O'Cléirigh, Micheál, et al. *Leabhar Gabhála: The Book of Conquests of Ireland.* Dublin: Hodges Figgis,1916.

O'Murchu, Diarmuid. *Quantum Theology.* New York: Crossroad, 1998.

Orr, David, W. *Earth in Mind.* Washington, DC: Island, 2004.

Otwell, John H. *And Sarah Laughed: The Status of Women in the Old Testament.* Philadelphia: Westminster, 1977.

Palese, Michela. "Which European Countries Use Proportional Representation?" Electoral Reform Society, Dec. 26, 2018. http://electoral-reform.org.uk/which-european-countries-use-proportional-representation/.

Pallotta, Nicole. "Federal Judge Strikes Down 'Lake Erie Bill of Rights.'" Animal Legal Defense Fund, May 4, 2020. https://aldf.org/article/federal-judge-strikes-down-lake-erie-bill-of-rights.

Parkinson, Hilary. "Sisters in Fate: The *Lusitania* and the *Titanic*." National Archives, May 1, 2012. https://prologue.blogs.archives.gov/2012/05/01/sisters-in-fate-the-lusitania-and-the-titanic/.

Parry, Hannah. "'There's Power in Love to Show Us the Way to Live': The Sermon in Full That Made Preacher Michael Curry the Breakout Star of the Royal Wedding." *Daily Mail*, May 19, 2018; last updated May 23, 2018. https://www.dailymail.co.uk/news/article-5747835/Bishop-raises-eyebrows-chapel.html.

Pennisi, Elizabeth. "Who Rules the World? Wild Mammals Far Outweighed by Humans and Domestic Animals." *Science*, Feb. 27, 2023. https://www.science.org/content/article/who-rules-earth-wild-mammals-far-outweighed-humans-and-domestic-animals.

Pilario, Daniel Franklin E., CM. "'Opening the Windows of the Church': Vatican II on the Church and the Modern World." *LiCAS*, Feb. 8, 2022. https://www.licas.news/2022/02/08/opening-the-windows-of-the-church-vatican-ii-on-the-church-and-the-modern-world/.

Pinker, Steven. *The Better Angels of Our Nature: Why Violence Has Declined*. London: Penguin, 2011.

———. *Enlightenment Now: The Case for Reason, Science, Humanism, and Progress*. New York: Penguin, 2019.

Pitt, David, and Paul R. Samson. *The Biosphere and Noosphere Reader: Global Environment, Society and Change*. London: Routledge, 1999.

Pratt Green, Fred. "The Church of Christ in Every Age." In *Hymns and Psalms: A Methodist and Ecumenical Hymn Book*, #804. London: MHP, 1983.

Rack, Henry D. *Reasonable Enthusiast: John Wesley and the Rise of Methodism*. 2nd ed. London: Epworth, 1990.

Ramis, Harold, dir. *Groundhog Day*. Culver City, CA: Columbia, 1993.

Raworth, Kate. *Doughnut Economics: Seven Ways to Think Like a 21st Century Economist*. White River Junction, VT: Chelsea Green, 2017.

Rees, William. "Dresden Nexus Conference 2015: William Rees—Keynote Speech." YouTube, Mar. 26, 2015. https://www.youtube.com/watch?v=icNvhnFv_wg.

Robertson, Alexander. "Fish and Chip Restaurant Closes." *Daily Mail*, July 9, 2019. https://www.dailymail.co.uk/news/article-7228331/Fish-chip-restaurant-closes-launch-new-gluten-free-plant-based-business-instead.

Roser, Max, and Hannah Ritchie. "How Has World Population Growth Changed Over Time?" Our World in Data, June 1, 2023. https://ourworldindata.org/population-growth-over-time.

Ruether, Rosemary. *New Woman, New Earth: Sexist Ideologies and Human Liberation*. New York: Seabury, 1975.

Samuel, Sigal. "Forget GDP—New Zealand Is Prioritizing Gross National Well-Being." *Vox*, June 8, 2019. https://www.vox.com/future-perfect/2019/6/8/18656710/new-zealand-wellbeing-budget-bhutan-happiness.

Savard, Jeff. *Labyrinths and Mazes: A Complete Guide to Magical Paths*. Asheville, NC: Lark, 2003.

Savard, Jeff, and Kimberly Savard. "Labyrinthos: Labyrinth & Maze Resource, Photo Library and Archive." https://www.labyrinthos.net/index.html.

Schreurs, Miranda, and Elim Papadakis. *Historical Dictionary of the Green Movement*. 3rd ed. Lanham, MD: Rowman & Littlefield, 2020.

Schweitzer, Albert. *The Quest of the Historical Jesus: A Critical Study of Its Progress from Reimarus to Wrede*. Translated by W. Montgomery. New York: Macmillan, 1956.
Schweizer, Richard. *Jesus*. Translated by David E. Green. London: SCM, 1971.
Scriven, Joseph Medlicott. "What a Friend We Have in Jesus." In *Methodist Hymn and Tune Book*, #332. Toronto: Methodist, 1917.
Sellar, W. C., and R. J. Yeatman. *1066 and All That: A Memorable History of England; Comprising All the Parts You Can Remember, Including One Hundred and Three Good Things, Five Bad Kings, and Two Genuine Dates*. 2nd ed. London: Methuen, 1949.
Sessions, George, ed. *Deep Ecology for the 21st Century: Readings on the Philosophy and Practice of the New Environmentalism*. Boston: Shambhala, 1995.
Shatner, William, and Josh Brandon. *Boldly Go: Reflections on a Life of Awe and Wonder*. New York: Atria, 2022.
Shaw, Graham. *God in Our Hands*. London: SCM, 1987.
Sheppard, David. *Bias to the Poor*. London: DLT, 1983.
Shore-Goss, Robert E. *God Is Green: An Eco-Spirituality of Incarnate Compassion*. Eugene, OR: Cascade, 2016.
Short, Robert L. *The Parables of Peanuts*. Glasgow: Collins Fontana, 1969.
Sierra Club Canada Foundation. "Oral Presentation." Canadian Nuclear Safety Commission, Apr. 7, 2022. CMD 22-H7.41. https://api.cnsc-ccsn.gc.ca/dms/digital-medias/cmd22-h7-41.pdf/object?subscription-key=3ff0910c6c54489abc34bc5b7d773be0.
Simard, Suzanne. *Finding the Mother Tree: Discovering the Wisdom of the Forest*. New York: Lane, 2021.
Slagle, Nicolette. "Rights of Nature FAQ." International Joint Commission, Oct. 2019. https://www.ijc.org/system/files/commentfiles/2019-10-Nicolette%20Slagle/FAQ.pdf.
Sloan Wilson, David. *This View of Life: Completing the Darwinian Revolution*. New York: Pantheon, 2019.
Souder, William. *Under a Wild Sky: John James Audubon and the Making of the Birds of America*. New York: North Point, 2004.
Spear, Stephen. "The Transformation of Enculturated Consciousness in the Teachings of Jesus." *Journal of Transformative Education* 3 (2005) 354–73.
Speller, Julian. *Galileo's Inquisition Trial Revisited*. Frankfurt: Lang, 2008.
Stein, Joseph. *Fiddler on the Roof*. New York: Pocket, 1965.
Stone, Christopher D. *Should Trees Have Standing? Towards Legal Rights for Natural Objects*. Palo Alto, CA: Tioga, 1988.
Straughan, Peter. *Wolf Hall*. Based on *Wolf Hall*, by Hilary Mantel. Aired Jan. 21–Feb. 25, 2015, on BBC Two.
Stuhlmacher, Peter. "The Messianic Son of Man: Jesus' Claim to Deity." In *The Historical Jesus in Recent Research*, edited by James D. G. Dunn and Scot McKnight, 324–34. Winona Lake, IN: Eisenbrauns, 2005.
Tallamy, Douglas W. *Nature's Best Hope: A New Approach to Conservation That Starts in Your Yard*. Portland: Timber, 2019.
Taylor, John V. *Enough Is Enough*. London: SCM-Canterbury, 1975.
Teilhard de Chardin, Pierre. *Activation of Energy*. Translated by René Hague. London: Collins, 1970.
———. *L'Avenir de l'homme*. Paris: Seuil, 1959.

———. *Le Phénomène humain*. Paris: Seuil, 1955.
———. *The Phenomenon of Man*. Translated by Bernard Wall. New York: Harper & Row, 1961.
Theissen, Gerd. *The Shadow of the Galilean: The Quest of the Historical Jesus in Narrative Form*. Translated by John Bowdon. London: SCM, 1987.
Tillich, Paul. *The Courage to Be*. New Haven, CT: Yale University Press, 1952.
———. *Dynamics of Faith*. New York: HarperOne, 1957.
———. *The New Being*. New York: Scribner's Sons, 1955.
Tupey, Marian L., and Gale Lyle Pooley. *Superabundance: The Story of Population Growth, Innovation, and Human Flourishing on an Infinitely Bountiful Planet*. Washington, DC: Cato Institute, 2022.
Tzaferis, Vassilios. "Jewish Tombs at and near Giv'at ha-Mivtar." *Israel Exploration Journal* 20 (1970) 18–32.
Uhl, Christopher. *Developing Ecological Consciousness: The End of Separation*. 2nd ed. Lanham, MD: Rowman and Littlefield, 2013.
UK Parliament. "Parliament Abolishes the Slave Trade." UK Parliament, n.d. https://www.parliament.uk/about/living-heritage/transformingsociety/tradeindustry/slavetrade/overview/parliament-abolishes-the-slave-trade.
UNFPA. *UNFPA Strategy for Family Planning, 2022–2030: Expanding Choices—Ensuring Rights in a Diverse and Changing World*. New York: United Nations Population Fund, 2022. https://www.unfpa.org/publications/unfpa-strategy-family-planning-2022-2030.
United Church of Canada, The. "A New Creed (1968)." UCC, 1968; last revised 1985. https://united-church.ca/community-and-faith/welcome-united-church-canada/faith-statements/new-creed-1968.
Van Houtan, Kyle S., and Michael S. Northcott, eds. *Diversity and Dominion: Dialogues in Ecology, Ethics, and Theology*. Eugene, OR: Cascade, 2010.
Van Montfoort, Trees. *Green Theology: An Eco-Feminist and Ecumenical Perspective*. Translated by Wim Reedijk. London: DLT, 2022.
Vega-Cárdenas, Yenny, and Nathalia Parra. "Nature as a Subject of Rights: A Mechanism to Achieve Environmental Justice in the Atrato River Case in Colombia." In *Extractive Industries and Human Rights in an Era of Global Justice: New Ways of Resolving and Preventing Conflicts*, edited by Amissi Manirabona and Yenny Vega-Cárdenas, 128–61. Toronto: LexisNexis Canada, 2019.
Vega-Cárdenas, Yenny, and Daniel Turp. *A Legal Personality for the St. Lawrence River and other Rivers of the World*. Montreal: JFD, 2023.
Verbeek, Peter, and Frans B. M. de Waal. "Primate Relationship with Nature: Biophilia as a General Pattern." In *Children and Nature: Psychological, Sociocultural, and Evolutionary Investigations*, edited by Peter H. Kahn Jr. and Stephen R. Kellert, 1–27. Cambridge, MA: MIT Press, 2002.
Vermes, Geza. *The Nativity: History and Legend*. London: Penguin, 2006.
Victor, Peter A. *Managing without Growth: Slower by Design, Not Disaster*. 2nd ed. Cheltenham, UK: Elgar, 2019.
Visser 't Hooft, Willem Adolf. "The Word 'Ecumenical'—Its History and Use." In *A History of the Ecumenical Movement, 1517–1948*, edited by Ruth Rouse et al., 729. London: SPCK, 1954.
Wagner, Thomas. "Social Tipping Point and Optimism." Bonpote, Apr. 18, 2020. https://bonpote.com/en/social-tipping-point-and-optimism/.

Waldfogel, Joel. *The Tyranny of the Market: Why You Can't Always Get What You Want.* Cambridge, MA: Harvard University Press, 2007.
Walker, Alice. *The Color Purple.* London: Women's, 1983.
Wang, Nanfu, and Lynn Zhang, dirs. *One Child Nation.* Santa Monica, CA: Amazon Studios, 2019.
Ward, Terry. "Is This Floating Eco-Pod the Future of Overwater Bungalows?" Waterstudio.NL, June 2, 2023. https://www.waterstudio.nl/is-this-floating-eco-pod-the-future-of-overwater-bungalows/.
Waxman, Rebecca, and Adam Meadows. "Blog: Day 1—Hearing for Radioactive Waste Dump Commences for Chalk River." Canadian Environmental Law Association, May 30, 2022. https://cela.ca/blog-hearing-for-radioactive-waste-dump-commences-for-chalk-river-laboratories-site.
Webster, Alison R. *Found Wanting: Women, Christianity and Sexuality.* New York: Cassell, 1995.
Weir, Peter, dir. *The Truman Show.* Los Angeles: Paramount, 1998.
Wesley, Charles. "Come, O Thou Traveller Unknown." In *Hymns and Psalms: A Methodist and Ecumenical Hymn Book,* #434. London: MHP, 1983.
Wesley, Cindy. "What Have the Sermons of John Wesley Ever Done for Us? John Wesley's Sermons and Methodist Doctrine." *Holiness: The Journal of Wesley House Cambridge* 1 (2015) 131–40. https://www.wesley.cam.ac.uk/wp-content/uploads/2015/05/10-wesley.pdf.
Wesley, John. *Forty-Four Sermons: Sermons on Several Occasions.* 1st ser. London: Epworth, 1944.
Westmount Mag. "Saving the Technoparc Wetlands." *Westmount Mag,* Mar. 29, 2017. https://www.westmountmag.ca/technoparc-wetlands/.
White, Lynn. "The Historical Roots of Our Ecological Crisis." *Science* 155 (1967) 1203–7.
Wilber, Ken. "Holons: Turtles All the Way Up, Turtles All the Way Down." Integral Life, Oct. 28, 2014. http://www.integrallife.com/holons-turtles-all-way-turtles-all-way-down/.
———. *Sex, Ecology, Spirituality: The Spirit of Evolution.* Rev. 2nd ed. Boston: Shambhala, 2000.
Williams, H. A. *The True Wilderness.* London: Constable, 1965.
Williams, Michael. *Deforesting the Earth: From Prehistory to Global Crisis.* Abridged. Chicago: University of Chicago Press, 2006.
Wilson, Edward O. *Biophilia.* Cambridge, MA: Harvard University Press, 1984.
———. *Half-Earth: Our Planet's Fight for Life.* New York: Liverlight, 2016.
Wink, Walter. *Engaging the Powers: Discernment and Resistance in a World of Domination.* Minneapolis: Augsburg Fortress, 1992.
———. *Transforming Bible Study: A Leaders Guide.* 2nd ed. Nashville: Abingdon, 1989.
Wohlleben, Peter. *The Hidden Life of Trees: What They Feel, How They Communicate.* Translated by Jane Billinghurst. Vancouver: Greystone, 2016.
Wolfe, Shira. "Stories of Iconic Artworks: Mark Rothko's Seagram Murals." *Artland Magazine,* n.d. https://magazine.artland.com/stories-of-iconic-artworks-mark-rothko-seagram-murals.
World Council of Churches. "About the WCC Logo." World Council of Churches, n.d. https://www.oikoumene.org/resources/logo.

Wright, Craig M. *The Maze and the Warrior: Symbols in Architecture, Theology, and Music.* Cambridge, MA: Harvard University Press, 2001.

Young, Frances, M. *Construing the Cross: Type, Sign, Symbol, Word, Action.* Didsbury Lectures 2014. Eugene, OR: Cascade, 2015.

Yunt, Jeremy D. *Faithful to Nature: Paul Tillich and the Spiritual Roots of Environmental Ethics.* Santa Barbara, CA: Barred Owl, 2017.

Zeffirelli, Franco, dir. *Jesus of Nazareth.* Rome: Rai Uno TV, 1977.

Index

Abram, David, 159
Adams, Douglas, 115
adult population groups, 98
Agapius, 23
Agnew, Paul, 56n37
"Ah, Holy Jesus" (hymn), 53
Alberta, Canada, 10
Alkali Act of 1863 (UK), 45
Allen, Woody, 43
Allison family, 8–9
Amazing Grace (film), 55
American Beauty (film), 110
Anadakat, Aatkin, 40
Anders, William, 171
Andersen, Hans Christian, 163
Anthropocene, 81, 107
anthropocentrism, 154, 169
Arcand, Denys, 133
Aristotle, 99
Attenborough, David, 3
Audubon, John James, 53–54
Augustine of Hippo, 30
autopoiesis (self-maintenance), 96
Azimi, Nassrine, 144

Bacon, Francis, 95
Baldwin, James Mark, 98, 100
Barth, Karl, 63
Barton, William, 156
Basilica of San Clemente al Laterano, (Rome), 185–86
Bauer, Bruno, 21
Bazzi, Giovanni Antonio "Sodoma," 140
Beatles, "Imagine" (pop group), 132–35

Besson, Luc, 34
Better Life Index, 175
The Big Blue (film), 34
biodiversity
 agricultural reforms, 178
 artificial reefs as, 179
 GDP and, 175
 for human diversity, 48
 of humanity, 29
 loss of, 56, 78
Biodiversity Conference, Montreal (2022), 160–61
biofeedback, 45, 175
biogenesis, 91
biophilia, energy as, 39–44
Biopsychosocial Model of Health, 85
biosphere, 80, 91, 187
birds, 1–2, 31, 34, 46
Birds of America (Audubon), 53–54
Birkenhead (ship), 166
bison, near extinction of, 47
Boff, Leonardo, 28
Boleyn, Ann, 18
Boswell, James, 29n18
Brown, Molly Young, 84
Buckingham, J. H., 166
Butterfly effect, 98

Cameron, James, 8, 165
Cantwell-Smith, Wilfred, 100
Capra, Fritjof, 80, 99, 114
Carson, Rachel, 79–80
cells (basic unit of life), 96–97, 128, 159
Celtic spirituality, 67
Charles II, King of England, 180

choice, in work and play, 39, 48–52
Christ (Messiah), 13, 23–24, 65–66, 128–29, 216
. *See also* Jesus
A Christmas Carol (Dickens), 186–87
chronos, *Kairos* and, 137–38, 144
Cleaver, Alison, 8
Clifton (ship), 167
climate change, 9–10, 165
coffee, dependency, 62
cognition, 97
"Come, O Thou Traveller Unknown" (hymn), 32–33
concentration camps, Germany, 182–83
consciousness
   dreams and nightmares, 52–57
   global, 13, 91–92
   as a green dimension, 39
   integral, 98–101
   new consciousness, 11–15, 93
   process of evolution, 93
COP15 Biodiversity Conference, Montreal (2022), 160–61
cosmogenesis, 91
Costa Concordia (ship), 9
Council of Nicaea (325), 19–20
councils, Ecumenical, 19–20, 89, 135–36
The Court of Nature, 191–93
COVID pandemic, 99, 105, 122, 173, 188
creation stories, 18–19
Cromwell, Thomas, 18
cross
   Greek for, 139
   home through, 179–189
   of Jesus Green, 180, 187–88
   new meanings from, 185
crucifixion, 140–42, 151–54
Crüger, Johann, 53
Crux Simplex (Lipsius), 140
Cullinan, Cormac, 169
Curie, Marie, 42
"currency of the commons," 103

da Vinci, Leonardo, 11, 95
Daedalus (Greek mythology), 34–35
Dali, Salvador, 185
Daly, Herman, 175
Dangerous Grounds (TV series), 62
Daniels, Bess Waldon, 8
Danzin, André, 93
Darwin, Charles, 41, 88, 93
De cruce (Lipsius), 140
de Wall, Frans B. M., 40–41
Deepwater Horizon disaster (2010), 172
Democritus (philosopher), 124
Descartes, René, 55, 93
Diatessaron (Tatian), 20
Dickens, Charles, 3, 27, 106, 187
Doc Martin (BBC comedy series), 101
Doughnut Economics (Raworth), 173
"A Dream of the Rood" (poem), 142–43
Dunn, James, 147
Dürer, Albrecht, 28
Dynamics of Faith (Tillich), 64

Earthrise photo, 178
Eco, Umberto, 20
ecology
   deep ecology, 95–96
   definition, 88
   history of experiment, 77–78
   Oikos Venn diagram, 87–88
   shallow ecology, 95
   trophic flow, 56, 177
economics, 87, 88
ecumenical, 87–89
   . *See also* councils, Ecumenical
Einstein, Albert, 42, 55
Eisenstein, Charles, 101–2, 104–5
Ekins, Paul, 175
electric car, 104
emergence, 128–131
emergent properties, 96
energy
   as biophilia, 40–44
   in evolution, 92
   Green, 39
   nuclear, 89–90
Enough Is Enough (Taylor), 79
environmental generational amnesia, 44

environmental justice theory, 169
environmental movements, 39, 45, 47
evolution, 93–94, 159, 176
extinction, 47, 54, 80–81, 165, 176, 191
Extinction Rebellion (global environmental movement), 47

faith
    as courage to be, 100–101
    dynamics of, 100
    existential, 100
    stages of, 100–101
The Five Gospels (Funk et al.), 20
Flanzraich, Aaron, 130
Foreign Slave Trade Abolition Bill of 1806 (UK), 55
Fowler, James, 100–101
fractals, 96
Francis, Pope, 136
Franciscan movement, 20
Friedman, Milton, 50
Friends of the Earth (organization), 40
Fromm, Erich, 40
Fukushima, Japan tsunami (2011), 90
Funk, Robert W., 20

Gaia/Gaia hypothesis, 94–95, 153, 171
Galileo Galilei, 95, 137
Gandhi, Mahatma, 14
Gardiner, Stephen, 18
GDP (Gross Domestic Product), 172–73
Genuine Progress Indicator, 175
geogenesis, 91
George, Henry, 103
Gesell, Silvio, 104
Ghost in the Machine (Koestler), 99
Gibson, Mel, 133
gift economy, 105
Global Greens Charter, 48, 48n23
Golding, William, 95
Goodall, Jane, 150
The Great Turning (Korton), 107–8, 131, 176
Green
    bible. *see* parables

consciousness, 97
consumer, 49–50
deeper, 77–82
dimensions of, 39
faint, 61, 73, 75, 108, 119
greenwashing, 39
movement, 39, 45
politics and, 44–48
Green Legacy Hiroshima (GLH), 144
GreenBiz, 49
Greenpeace, 47–48
Gross Domestic Product (GDP), 172–73
Gross National Happiness, 175
Groundhog Day (film), 64
growth
    limits to, 159–160
    managing without, 173–74
    of population, 81
Gutiérrez, Gustavo, 28

Haeckel, Ernst, 88
Haitian uprising (1805), 55
Hammerstein, Oscar, 52
Happy Planet Index, 175
Helena, Queen (Constantine), 180–81
Henry, Patrick, 63–64
Henry VIII, King of England, 18
Hildegard of Bingen, 26
Hiroshima, Japan, 143–44
historical thinking, 63–64
Hitchhiker's Guide to the Galaxy (Adams), 115
Hockney, David, 36
holonic, 128
holons/holonic theory, 98–99, 128
home
    for all, 171–79
    homeless, 3, 88
    Odysseus, 163–65
    through the cross, 179–189
Honest to God (Robinson), 136
Hoover, Roy W., 20
Hopkins, Gerald Manley, 3
human costs, to rapid social change, 10–11
Human Development Index (HDI), 175

The Human Phenomenon (Huxley), 91
human-divine relationships, 121–22
Huxley, Julian, 90–91

Icarus (Greek mythology), 34
"Imagine" (song), 132–35
interaction, with the environment, 96–97
International Observatory on the Rights of Nature, 169
Irenaeus, Bishop, 65
"I-Thou" experience, 115–17, 125

James Webb Space Telescope, 93
Jesus
    anointing of at Simon's house, 117–19
    Arctic world, 21
    the chameleon, 29–34
    the disturber, 23–25
    in films, 133–34, 135
    hand of, 183–84
    as homeless, 168
    many versions of, 16–21, 129
    names for, 145–46. see also Christ (Messiah)
    of Nazareth, 13, 119, 145–151, 153
    poor, biased to, 27–28
    as pro-women, 26–27
    renamed, 109–14, 128–29
    same-sex love bias, 28–29
    "second coming," 165
    as systems thinking, 119–124
    true vine, 185–86
Jesus, the True Vine (painting), 186–87
Jesus Green, 1–3, 7, 12–15, 34–37, 128–131
Jesus of Montreal (film), 133, 135
Jesus of Nazareth (TV film), 29
Jewish Antiquities (Josephus), 67–68
John, of Patmos, 134
John of the Cross, Saint, 185
John the Baptist, 67–68
John XXII, Pope, 20
John XXIII, Pope, 135
Josephus, Flavius, 24, 66, 67–68, 140–41
Julian of Norwich, 26

Kairos in Soho (London), 35, 137–38, 144
Kelsey, Charles W., 6
kingdom of God/Heaven. See realm
Kipling, Rudyard, 166
The Kiss (Rodin), 130
Koestler, Arthur, 99
Korton, David, 107
Koyama, Kosuke, 183–84
Kramer, Jeff, 36

labyrinths, 20, 35–37
Lake Baikal, Siberia (2021), 173
Lake Erie Bill of Rights of 2019 (LEBOR), 169
Last Supper (da Vinci), 11
László, Ervin, 98, 99, 101
Laudato Si' (papal encyclical), 136
Le phenomène humain (Teilhard), 176
Leaves of Grass (Whitman), 3
Lederman, Leon, 123–24
left-and right-handedness, 139
Lennon, John, 132–33
Leopold, Aldo, 56
Lewis, John, 193
LGBTQ+ community, 35–36, 80, 137–38
    . See also same-sex love bias
life
    characteristics, 159–163
    definition, 96
    history of, 18, 91
Life of Brian (film), 50
limitations, in life, 159–163
The Limits to Growth (Club of Rome), 46
Linnaeus, Carolus "Carl," 41, 88
Lipsius, Justus, 140
Logan, Joshua, 52
London (UK), 35, 137–38, 144
love
    commandment to, 6, 70, 72, 112, 151
    of living things (biophilia), 40–44, 153
    of nature, 6, 67, 77–78, 157
    same-sex love bias, 28–29

INDEX

Love in the Time of Cholera (Márquez), 1
Lovelock, James, 94–95
Luisi, Pier Luigi, 80, 99
Lusitania (ship), 167
Luther, Martin, 63–64

Macy, Joanna, 84, 107–8
Magpie River (Romaine River), 169–170
Makower, Joel, 49–50
Malina, Bruce, 26–27
Malthus, Thomas, 34
Managing Without Growth (Victor), 173
mandrill monkeys, 41
Marcion (gnostic Christian leader), 65
Margulis, Lynne, 94
Márquez, Gabriel García, 1
massage, 36, 58
Masurel, Jacques, 93
Matthew, Gospel of, 13–14
mazes, 35
McCartney, Paul, 133
McFague, Sallie, 151
McKintosh, Steve, 98
McRobie, George, 175
Merrily We Roll Along (musical), 106
Messiah (Christ), 13, 23–24, 65–66, 128–29, 216
  . *See also* Jesus
Messiah (Handel), 51
Metaphysics (Aristotle), 99
Methodist Church, 83, 120, 132
A Midsummer Night's Dream (Shakespeare), 46
Migratory Bird Conservation Act (1929), 45
mind
  "crucified mind," 183–84
  noogenesis, 91, 176, 178
  oikogenesis, 179
money
  alternative currencies, 103–5
  biblical references, 105–7
  currency of the commons, 103
  history of, 101–3
Montreal, Canada, 1–2, 160–61

Monty Python, 50
Moore, Sebastian, 188
movements
  environmental, 39, 45, 47
  politics and, 39, 44–48
  social credit, 105
Muir, John, 47, 67

Naess, Arne, 80–81, 95
The Name of the Rose (film), 20, 35
names, changing of, 109–14, 128–29
Natal Game Preservation Society, 45
National Trust of 1895 (UK), 45
nature
  abuse of, 43n11
  fear of, 42, 47, 56
  love for, 6, 67, 77–78, 157
  respect for, 42, 57, 87, 153, 169
  rights of, 169
near-surface disposal facility (NSDF), 89
Neef, Manfred Max, 175
Neil, Clarence, 40–41
Neusner, Jacob, 65
The New Being (Tillich), 138
New Economics Foundation, 175
Newton, Isaac, 41, 55
Nicene Creed, 20
Niebuhr, Richard, 100
No Handle on the Cross (Koyama), 183
non-localization, 96
noogenesis, 91, 176
noosphere, 91, 178
nuclear waste, 90

Odysseus, 164
oikogenesis, 178–79, 184
oikos, 82–90, 174
Oikos Venn diagram, 87–88
oikosphere, 178
Old Plantation Hymns (Burton), 156
orangutan, 138, 154, 156, 191
Organization for Economic Cooperation and Development (OECD), 175
The Origins of Species (Darwin), 3, 88
Orr, David, 43

overshoot, 40, 44, 57, 144, 154
Oz (TV series), 35

Paine, Robert, 77–78
palm oil, 43, 154
parables
   characteristics of, 119–120
   Green examples, 13, 60, 73, 110, 120–21
   house building, 78
   human-divine relationships, 121–22
   mode of thought, 23–24, 68, 71
paradox, 73
People, Profit, Planet (Venn diagram), 86
Pereira, Fernando, 48
Perry, Alexander, 5, 6
Peter Getting Out of Nick's Pool (painting), 36
The Phenomenon of Man (Teilhard), 45, 176
Piaget, Jean, 98, 100
Picasso, 71
Pinker, Stephen, 55
Poland (packet boat), 166–67
political movement of Green, 39, 44–48
politics, of Gospel in different regions, 20
pollution
   currency of the commons and, 103
   Doughnut Economics and, 173
   of "fast fashion," 49
   Leicester fish, 40
   light pollution, 178
   noise pollution, 174
   of water, 75, 170, 172
popularism, 4
population
   control of, 81
   growth of, 10, 44–45, 107
   stages of development, 98, 162
Powell, Robert, 29
predator/predation, 78
prey, 33, 54, 78, 177

Raworth, Kate, 173, 175
Ray, Paul, 98
realm, 13–14, 59–60, 73n14, 121–22, 152
reconciliation
   of Christianity and science, 90
   the cross and, 185, 187
   of God to humanity and all of creation, 153
   with nature, 188
   need of, 37
   new reconciliation, 189
   process of, 177
Rees, William, 39n1
rights for nature, 169
Robinson, John T., 136
rock pool, 77–79
Rodin, Auguste, 130
Rogers, Richard, 52
Romaine River (Magpie River), 169–170
Romeo and Juliet (Shakespeare), 155
Rothko, Mark, 116
Rothko exhibition (Britain), 116
Rowland-Molina hypothesis (1974), 46n19

Sacred Economics (Eisenstein), 101, 105
same-sex love bias, 28–29
   . *See also* LGBTQ+ community
Schumacher, E. F., 175
Schweitzer, Albert, 20–21
science, spirit/spiritual and, 90–94, 158–59
Scripture
   analysis of, 18–19, 62–64
   biblical language and meaning, 66–70
   continuity and discontinuity, 65–66
   Green references, 59–61
   on money, 105–7
   originals, 19–20, 24
   unity and diversity of, 65
The Seagram Murals (Rothko), 116
Sebastien, Saint, 140
self-maintenance (autopoiesis), 96

Sessions, George, 80–81
sex worker
  female, 119
  male, 164
Shakespeare, William, 31, 46, 71, 155
Shatner, William, 171, 178
shipwrecks
  Birkenhead, 166
  Costa Concordia, 9
  Lusitania, 167
  Poland, 166–67
  Titanic, 8–11, 165–66
Short, Robert, 71
Sierra Club, 89–90, 89n20
Silent Spring (Carson), 79–80
Simard, Suzanne, 92
slavery, 55, 156
Smith, Edward J., 9
social change
  human costs, 10
  slavery, abolishment of, 55
  timeline, 161
  tipping point, 75, 97, 99, 101
social credit movement, 105
Soho, 8, 35, 137–38, 144
Soil Conservation Society of America, 45
Sondheim, Stephen, 106
song and singing, 31–33
South Pacific (musical), 52
Star Trek (TV series), 171
Stone, Christopher, 169
Strauss, David, 20
symbols, 130–31, 135
systems thinking, 94–101, 119–124
The Systems View of Life (Capra & Luisi), 80

The Tao of Physics (Capra), 114
Technoparc wetland (2017), 151
Teilhard de Chardin, Pierre, 45, 90–94, 98, 175–76
Temple, (Second), Father's House, 124–28
Thatcher, Margaret, 28
Theissen, Gerd, 23
Thoreau, Henry David, 47
Thunberg, Greta, 90, 95, 97, 162–63

Tillich, Paul, 31, 64, 99–100, 114, 130–31, 138, 182–83
Titanic (film), 165
Titanic (ship), 8–11, 165–66
trees
  biblical respect for, 60–61, 141–42
  communication between, 92
  in Greek, 141
  legal rights of, 169
  symbol for, 92
The Truman Show (film), 110
The Tudors (TV series), 18
Turin Shroud, 13
Tzaferis, Vassilios, 139

Uffizi art gallery (Florence, Italy), 139–140
United Church of Canada, 6
United Nations, 81, 97, 160–63, 175
Universal Declaration of the Rights of Mother Earth (2010), 169
unprecedented, term usage, 81

Vatican II (second-ever papal council), 135–36
Vega-Cárdenas, Yenny, 169
Venn, John, 84
Venn diagrams
  Biopsychosocial Model of Health, 85
  Oikos Venn diagram, 87–88
  People, Profit, Planet, 86
Verbeek, Peter, 40–41
Vernadsky, Vladimir, 176
Victor, Peter A., 173–74

Wade, Jeremy, 42
Waldfogel, Joel, 50
Walker, Alice, 3
Watanabe, Tomoko, 144
Water Buffalo Theology (Koyama), 183
"Were You There" (hymn), 156
Wesley, Charles, 32–33, 120
Wesley, John, 39, 48, 120
Wesleyan Quadrilateral, 39
Westmount Park United Church, 4–8, 10–11, 67, 105, 144, 145
White, Lynn, 47

Whitefield, George, 120
Whitehead, Alfred, 92
Whitman, Walt, 3, 150
Wilber, Ken, 98–99, 128
wild law initiatives, 169
wilderness
   formation, 67–70
   renewal, 56
   spirituality, 67–70
Williams H. A., 115n10
Wilson, David Sloan, 45
Wilson, Edward O., 40
Wink, Walter, 92, 123n21
Wohlleben, Peter, 92
Wolf Hall (book, TV series), 18
women and children, priority, 165–171
Wordsworth, William, 3
World Council of Churches (WCC), 88–89

"you-me-us" experience, 35, 116

www.ingramcontent.com/pod-product-compliance
Lightning Source LLC
Chambersburg PA
CBHW070316230426
43663CB00011B/2147